THE STRATEGIC DIMENSION
OF MILITARY MANPOWER

THE STRATEGIC DIMENSION OF MILITARY MANPOWER

Edited by
GREGORY D. FOSTER
ALAN NED SABROSKY
WILLIAM J. TAYLOR, JR.

With a Foreword by Robert C. McFarlane

BALLINGER PUBLISHING COMPANY
Cambridge, Massachusetts
A Subsidiary of Harper & Row, Publishers, Inc.

Copyright © 1987 by Center for Strategic and International Studies. All rights reserved. No part of this publication may be reproduced, stored in a retrieval system, or transmitted in any form or by any means, electronic, mechanical, photocopy, recording or otherwise, without the prior written consent of the publisher.

International Standard Book Number: 0-88730-155-X

Library of Congress Catalog Card Number: 86-28764

Printed in the United States of America

Library of Congress Cataloging-in-Publication Data

The strategic dimension of military manpower.

   Papers presented at a conference held in May 1985 at the Georgetown University Center for Strategic and International Studies in Washington, D.C.
   Includes index.
   1. Manpower—Congresses. 2. Manpower—United States—Congresses. I. Foster, Gregory D. II. Sabrosky, Alan Ned, 1941–   III. Taylor, William J. (William Jesse), 1933–
IV. Georgetown University. Center for Strategic and International Studies.
UA17.5.A2S77 1987   355.2′2   86-28764
ISBN 0-88730-155-X

CONTENTS

List of Figures and Tables	ix
Foreword—*Robert C. McFarlane*	xi
Preface	xiii

PART I   *INTRODUCTION*

Chapter 1
Overview
—*Gregory D. Foster and William J. Taylor, Jr.*   3

PART II   *THE SETTING*

Chapter 2
Manpower as an Element of
Military Power
—*Gregory D. Foster*   13

Chapter 3
The Role of Manpower in Traditional
Strategic Thought
—*John Keegan*   37

PART III THE CONTEMPORARY SITUATION

Chapter 4
Military Manpower in Current U.S. Strategic Planning
—Robert B. Pirie, Jr. 53

Chapter 5
Strategic Influences in Military Manpower Planning
—Sam C. Sarkesian 65

PART IV MILITARY MANPOWER OPTIONS

Chapter 6
Manpower Procurement Options: The Influence of Demography, Technology, and Budgets
—Martin Binkin 91

Chapter 7
Manpower Procurement and Military Doctrine: "You Can't Get There From Here"
—William L. Hauser 117

PART V PROSPECTS FOR A FUTURE SYNTHESIS

Chapter 8
Implications of Likely Future Conflict Environments for U.S. Military Manpower Policies and Practices
—Jeffrey Record 143

Chapter 9
Human Resources and Military Requirements: Strategic Considerations on Past and Future
—Irving Louis Horowitz 163

PART VI   ROUNDTABLE DISCUSSIONS

Chapter 10
Manpower and Strategy: Issues in Methodology and Analysis
—*George W. Sinks*   185

Chapter 11
Cross-National Assessments of the Manpower-Strategy Interface
—*Karen A. McPherson*   193

PART VII   SUMMARY AND CONCLUSIONS

Chapter 12
The Strategy-Manpower Interface: Retrospect and Prospect
—*Alan Ned Sabrosky and
   William J. Taylor, Jr.*   211

Index   225

About the Editors   237

About the Contributors   239

LIST OF FIGURES AND TABLES

Figures
5-1 The Conflict Spectrum   75
5-2 The Conflict Spectrum—Revised   76

Tables
6-1 Percentage Distribution of Recruits, by Aptitude Category and Level of Education, All Services, Selected Fiscal Years, 1960-1984   94
6-2 Percentage Distribution of Army Recruits, by Aptitude Category and Level of Education, Selected Fiscal Years, 1960-1984   95
6-3 Percentage of Males, Ages 16-21, with a Positive Propensity for Military Service, by Service of Choice, 1979-1984   96
6-4 Projected U.S. Population Aged 18 to 21, by Sex and Race, Selected Years, 1981-1985, in Thousands   98
6-5 Proportion of Qualified and Available Males Required for Military Service, Selected Periods, 1984-1995   99
6-6 Distribution of Trained Military Enlisted Personnel by Major Occupational Category, Selected Years   101
6-7 Tank Crew Performance   104

6-8	Percentage of American Youths (18-23 Years) Who Would Qualify for Selected Occupations in the Army and Navy (Arranged According to Occupational Area), by Sex and Racial/Ethnic Group	106
12-1	Architecture of Basic Force Structures	217

FOREWORD

Outside the Pentagon, defense manpower issues have not received priority for several years. The explanation for such a low priority is straightforward: Few military manpower problems have *required* attention. In the early 1980s the all-volunteer force proved to be a success largely because of the recession; military recruiting was aided by high unemployment and the lack of competitive job opportunities in the civilian sector. During that time, defense spending grew by almost 50 percent in real terms over five years, which permitted significant military pay increases, expanded benefits, increased training, and large investments in Madison Avenue-style recruiting efforts. Each of these significant factors improved the morale and bettered the lifestyle of military personnel and their families, which weighed heavily on reenlistment decisions.

The horizon is darkening, however. Improvements in the U.S. economy in general, coupled with the growth in attractive civilian job options, create new problems for military manpower planning. As congressional preoccupation with deficit reduction begins to affect the defense budget, incentives to remain in military service are almost certain to diminish. In addition, demographic trends indicate that the cohort of draft-age males will continue to decline through the early 1990s, creating a smaller manpower pool.

These trends indicate that over the coming two to three years crucial decisions affecting strategy, manpower, missions, training, force structure, and equipment must be made. Given declining budgets for conventional military forces, it must be decided what the proper strategy-manpower relationship should be. Stated simply, either a diminished military manpower pool must be dealt with directly or changes in present manpower policies given more serious consideration. Resolution of this dilemma will require the best efforts of our strategic and manpower planners. In a departure from past practice, these two groups must work closely together.

The contributors to this book are representative of the disciplines that relate to manpower issues, and their chapters provide valuable insights that strategic and manpower planners ignore at their peril, particularly during a time when defense resources for manpower are diminishing.

Robert C. McFarlane

PREFACE

The original idea for this project surfaced in 1981, following one of the periodic gatherings of the Inter-University Seminar on Armed Forces and Society. At that time, the new Reagan administration was sending strong signals that, despite what might be said about the efficacy of the All-Volunteer Force, the team at the Defense Department intended to make the idea work.

On the other hand, a flawed strategic performance by the United States in Vietnam and the less than successful prosecution of other more recent and limited military engagements—Son Tay, Mayaguez, and Desert One, to name but a few—had made it obvious that something had to be done about a serious conceptual and operational mismatch. Clearly, neither strategic planning nor military manpower policy adequately appreciated the other—this despite the fact that the importance of the human dimension of strategy and the strategic consequences of manpower are readily apparent to any serious student of military affairs.

The idea of gathering together a group of respected authorities on manpower and strategy for an exchange of ideas continued to germinate during the ensuing three years. The editors of this volume carried on a running dialogue with other like-minded individuals, concerned over the lack of communication in this area. Accordingly, plans were developed to convene such a gathering.

The result was a two-day conference, held in May 1985 at the Georgetown University Center for Strategic and International Studies in Washington, D.C. This book represents the product of that seminal gathering.

The editors wish to acknowledge the contributions of a number of individuals without whose efforts this endeavor would not have been possible. We are grateful, first, to Dr. Enid C.B. Schoettle, Ford Foundation program officer, whose support and guidance were indispensable to the success of this undertaking.

Second, thanks also are due the two individuals who, along with the editors, constituted the steering group for this venture: Dr. John Steinbruner, Director of the Foreign Policy Studies Program at The Brookings Institution, and Dr. Samuel F. Wells, Jr., Associate Director of the Woodrow Wilson International Center for Scholars and Director of the Wilson Center's International Security Studies Program. Both contributed their insights to the planning and conduct of the conference, and their affiliations lent the project a multi-institutional source of perspectives.

Third, the editors extend their appreciation to all of the conference participants, whose contributions produced a quality of discourse that, in every sense, measured up to the expectations of the conference organizers. In addition to those otherwise identified as contributors to this volume, special thanks are due the following people: The Honorable William K. Brehm, Shiela Brammer, Mr. Paul M. Cole, Professor Catherine Kelleher, The Honorable Amos A. Jordan, The Honorable Lawrence J. Korb, Dr. Franklin D. Margiotta, Roger Merritt, Professor Charles C. Moskos, Mrs. Jean Newsom, Professor David R. Segal, and Adam Yarmolinsky, Esq.

Finally, we extend our appreciation and respect to Morris Janowitz, Lawrence A. Kimpton Distinguished Service Professor at the University of Chicago, who inspired this effort. Professor Janowitz has provided intellectual leadership to scholars and practitioners of civil-military affairs for some four decades. His role in siring, nurturing, and legitimizing the study of armed forces and society has been singular and pervasive. An intellectual beacon amidst the seeming fog of contemporary military thought, he has been a source of inspiration and insight to many. More importantly, he has been mentor, colleague, and friend to an entire new generation of thinkers steeped in the sophistication of his views. The ultimate reach of his influence, however, is yet to be recorded on the minds of generations to come. This book, like the conference that spawned it, is therefore dedicated to Professor Janowitz, not merely to acknowledge his many accomplishments but to accord appreciation and respect for his profound intellectual leadership.

1 INTRODUCTION

1 OVERVIEW

Gregory D. Foster
William J. Taylor, Jr.

The recent past is always the surest guide to the future. By this measure alone, the years immediately ahead promise a period of intensified public and congressional scrutiny of the U.S. defense posture and, more generally, of this country's overall strategic direction. Assumptions and enduring truths left essentially unquestioned for almost four decades will undergo reconsideration. Heightened concern will be voiced over the relatively low marginal rate of strategic return on each dollar spent for defense, while the lack of "strategic sense" repeatedly demonstrated by the United States abroad will stimulate the reappraisal of issues previously given little regard in terms of their strategic ramifications.

In view of its importance to the American public, military manpower seems likely to become the object of especially close scrutiny. Although a great deal of excellent analysis has been devoted to manpower issues in recent years, the approach has generally been more actuarial than strategic in focus. But, just as manpower analysts have tended to ignore questions of larger strategic import, so-called strategic thinkers have treated manpower issues with considerable disdain. Thus, a seemingly natural conceptual union between manpower and strategy has, largely for reasons of academic and bureaucratic convenience, fallen prey to an artificial estrangement.

Recognizing this unfortunate state of affairs, the Georgetown Center for Strategic and International Studies in May of 1985 convened a conference entitled The Strategic Dimension of Military Manpower. Supported by a grant from the Ford Foundation, this gathering drew together leading authorities on manpower and strategy from the United States and abroad, and thereby stimulated a much-needed dialogue between these two communities. Of the some 120 scholars and practitioners who participated, nearly all entered into the proceedings with at least an intuitive sense that something was amiss and that this venture might help to illuminate the situation.

The papers presented in this volume provided the stimulus—and now represent the fruits—of that initial dialogue. They give no easy answers. Rather, their value lies in the questions they pose—questions that, to date, have not been brought into adequate focus. Hopefully, the questions themselves will provide an impetus for reform and bring manpower and strategy into closer conceptual and operational harmony, thus ensuring a more robust national security posture.

ORGANIZATION OF THE BOOK

This book attempts to set conceptual bounds on the strategy–manpower relationship. With few exceptions, chapters are paired so as to address particular dimensions of the problem from the independent perspectives of strategy and manpower. In this manner, it is possible to obtain a general sense of where areas of natural convergence and divergence exist.

Chapters 2 and 3, authored respectively by Gregory D. Foster and John Keegan, establish the overall setting for the chapters that follow. In Chapter 2, "Manpower as an Element of Military Power," Foster seeks to lay a conceptual foundation for linking manpower and strategy. He addresses himself to the following types of questions: What is the nature of power, and how does its application contribute to the effective formulation and execution of strategy? What characteristics make manpower an integral and unique element of military power? How can strategy best exploit these unique characteristics? What is the relationship of manpower to technology, doctrine, and force posture, and what is the strategic value of these elements, either singly or in combination? To the extent that deterrence and combat effectiveness can be differentiated in terms of their constituent elements, are there particular manpower considerations that contribute more to one than to the other? What are the potential strategic ramifications of such issues as elitism, conscription versus voluntarism, the active-reserve composition of the force, the light-heavy mix of the force, the role of females, and so on?

In Chapter 3, "The Role of Manpower in Traditional Strategic Thought," Keegan assays the significance, or insignificance, of manpower—and, more generally, of the human element—in classical and contemporary strategic thought. He addresses the following types of questions: Does manpower play a central or a peripheral role in the writings of leading strategic thinkers, past and present? What specifically do these individuals say about manpower? What does their emphasis, or neglect, suggest about the nature of the manpower-strategy relationship today? Were there unique historical or cultural factors that affected their views on manpower? Do the manpower-related views of leading strategic thinkers represent a major strength or weakness in the established corpus of strategic thought?

Chapters 4 and 5, authored respectively by Robert B. Pirie, Jr. and Sam C. Sarkesian, deal with the general nature of the contemporary planning process. In Chapter 4, "Military Manpower in Current U.S. Strategic Planning," Pirie addresses the role manpower plays in current strategic planning. He attempts to come to grips with the following types of issues: Does the nature of the strategic planning process provide for appropriate consideration of manpower issues? Why or not? What questions do strategic planners ask, and do these questions adequately accommodate the concerns of manpower planners? How do strategists view manpower experts? Does the current orientation in this regard differ materially from past orientations? From the standpoint of the strategic planner, what must be done and what questions asked to forge a more effective convergence of manpower and strategy? Is such a convergence deemed desirable?

In Chapter 5, "Strategic Influences in Military Manpower Planning," Sarkesian deals with the obverse of the above: the role strategic considerations play in current U.S. manpower planning. Sarkesian addresses the following types of questions: Does the nature of the manpower planning process provide for appropriate consideration of broad-gauged strategic issues? Why or why not? What questions do manpower planners ask, and do these questions adequately accommodate the concerns of strategic planners? How do manpower experts view strategists? Does the current orientation in this regard differ materially from past orientations? From the standpoint of the manpower planner, what must be done and what questions asked to forge a more effective convergence of manpower and strategy? Is such a convergence deemed desirable?

Chapters 6 and 7, authored respectively by Martin Binkin and William L. Hauser, focus on current military manpower options within their broader strategic context. In Chapter 6, "Manpower Procurement Options: The Influence of Demography, Technology, and Budgets," Binkin discusses the relative influence of such critical factors as demography, technology, and budgets in the determination of U.S. military manpower options. Binkin deals with the following types of questions: Are there discernible demographic, technological, and/or budgetary developments that promise to shape the composition and eventual selection of particular manpower alternatives in the years ahead? Which of these factors seems likely to predominate? How may these factors influence such questions as the desirability or undesirability of reinstituting conscription; the centrality of the reserve components in the total force posture; and the male–female, careerist–short-term, and age-distribution mixes constituting the force? In the final analysis, do such factors, either singly or in combination, exert a constraining influence, or do they facilitate full exploitation of manpower as an element of strategy? Most importantly, how significant an effect do (and will) fluctuations in the domestic economy have on the selection of manpower options?

In Chapter 7, "Manpower Procurement and Military Doctrine: 'You Can't Get There From Here,' " Hauser focuses on the potential consequences of various manpower options for U.S. military doctrine and force posture. Hauser confronts the following types of questions: What is, and should be, the proper relationship between the demographic and attitudinal characteristics of the American military on the one hand and, on the other, such fundamental questions as U.S. interventionist propensities, the assertiveness of military doctrine (e.g., warfighting/-winning vs. deterrence), nuclear reliance, and overall force orientation (e.g., offensive vs. defensive, forward-based vs. homeland-based)? Are U.S. doctrine and force posture sufficiently flexible to adapt readily to changing manpower realities? Why or why not? To the extent that fundamental attributes of the American character are reflected in the U.S. military, how appropriate and valid are reformist criticisms of the military's overreliance on attrition-oriented doctrine and the consequent call for a maneuver-based reorientation?

Chapters 8 and 9, authored respectively by Jeffrey Record and Irving Louis Horowitz, look at the prospects for a future synthesis of manpower and strategy. In Chapter 8, "Implications of Likely Future Conflict Environments for U.S. Military Manpower Policies and Practices," Record focuses on the future of conflict over the next two to three decades and attempts to discern the implications of that future for military manpower. He addresses the following types of questions: What will be the predominant nature and form(s) of conflict in the years ahead? What role will manpower play—especially vis-à-vis technology—in this conflict environment? What are the principal manpower attributes and capabilities that will be required? To the extent that multiple conflict forms can be expected to predominate, is there a consistency to the manpower capabilities that will be required? How relevant will the traditional, idealized soldierly virtues be to such a conflict environment? Do future conflict forms demand changes in prevailing U.S. attitudes toward such things as elitism, conscription, and military age and gender considerations? What directions must U.S. planning take if the strategic dimension of military manpower is to be exploited to its fullest in the face of the anticipated conflict environment? Will there be discernible geographic patterns to future conflicts (e.g., Europe vs. Southwest Asia vs. Latin America) that will materially affect U.S. reliance on and utilization of manpower?

In Chapter 9, "Human Resources and Military Requirements: Strategic Considerations on Past and Future," Horowitz looks at emerging manpower trends over the next two to three decades and attempts to discern the implications of these trends for U.S. strategy. He deals with the following types of questions: What seem likely to be the predominant demographic and attitudinal characteristics of the American populace in the years ahead? What do such factors portend for such questions as the utility of force as an instrument of national power, the propriety of intervention, the role of the military as a social institution, and the acceptability of sacrifice? Do these characteristics seem likely to exert any kind of inhibiting effect on the United States and its military in coping with an increasingly complex and demanding international environment? What adaptations, if any, must the United States make in its international behavior to accommodate these manpower trends more effectively? What will be the nature and significance of the domestic politics likely to surround any future effort to reinstitute conscription in the United States?

Chapters 10 and 11, authored respectively by George W. Sinks and Karen A. McPherson, summarize two roundtable discussions conducted during the course of the conference. Sinks discusses the roundtable that addressed the subject, "Manpower and Strategy: Issues in Methodology and Analysis," a forum for identifying points of convergence and divergence across academic disciplines involved in the study of manpower issues. Viewpoints representing the fields of history, sociology/social psychology, political science, and economics/econometrics are presented. From the particular perspective of the discipline represented, each participant addresses the following types of questions: What are the major manpower issues/themes dealt with by your discipline? Why? What are the fundamental assumptions underlying this focus? How does (and could) the focus of your discipline contribute to broader questions of military and national strategy? Do the tools and techniques employed by your discipline limit the types of issues that can be addressed and/or the quality of insights that can be derived? Juxtaposed against the views of the various academic disciplines are the views of a senior military practitioner, who assesses the effectiveness with which the military and the analytical community deal with each other on critical manpower issues.

In Chapter 11, McPherson discusses the roundtable that addressed the subject, "Cross-National Assessments of the Manpower-Strategy Interface." This roundtable sought to place the American approach to manpower and strategy in a broader international context by providing a sample of cross-national perspectives on the nature and importance of the manpower-strategy relationship. Viewpoints representing the USSR, two European countries (West Germany and France), and selected areas of the so-called Third World are presented. From the particular perspective of the country(ies) represented, each participant addressed the following types of questions: How important is manpower as an element of overall military strategy? What is the nature of the manpower-strategy relationship? Are there anticipated future developments/trends that promise to alter this relationship? Are there unique sociocultural, political, and/or ideological propensities that affect this orientation? What lessons are potentially applicable to the United States?

As the reader will quickly discern, each author has been given wide latitude in addressing the questions posed. Consequently, some of the questions delineated above may be adjudged less relevant than others that might have been addressed, while yet other questions not identified here will be seen to emerge during the course of the author's discussion. In Chapter 12, Alan Ned Sabrosky and William J. Taylor, Jr., synthesize the arguments put forth in this book and offer a speculative appraisal of the prospects for meaningful change in the years ahead.

II THE SETTING

2 MANPOWER AS AN ELEMENT OF MILITARY POWER

Gregory D. Foster

Even in this age of technological dominance, it is hard to quarrel with the intuitive appeal of Ardant du Picq's observation that "man is the foremost instrument of combat." Unfortunately, the blurring effects of contemporary conflict on the definitional limits of *combat* have made it extremely difficult to ascertain the proper role of military manpower in strategy.

Where once the human element occupied a position of clear primacy in military affairs, its position today is singularly ambiguous. Contemporary strategic planning—and, more broadly speaking, strategic discourse in general—evinces at best only a casual concern with the seemingly pedestrian domain of manpower. Manpower analysis, in turn, has contributed to its own strategic neglect by focusing on a range of problems sufficiently unsexy, and even obscure, as to fail to capture the hearts and minds of strategists.

Such dissociative tendencies derive from an admixture of reasons. For one thing, strategic studies over the past three and one-half decades have been the analytical preserve of a genre of nuclear-dominated issues and concepts—arms control, crisis management, deterrence, escalation control, and the like—to which manpower has been adjudged largely irrelevant. Similarly, both our overriding obsession with technology and our exceptional reverence for the value of human life have imbued us with a hopeful sense of evolutionary determinism that, implicitly at least, views technological advances as eventually rendering the human element in warfare obsolete. Finally, the centrality of budgetary bean counting in our defense decisionmaking process, reinforced by the false allure of scientific method embodied in systems analysis, has lent more than a little credence to Edward Luttwak's characterization of the American style of warfare as "war in the administrative manner."[1]

Collectively, these factors have produced a conceptual void that must be filled if manpower and strategy are to be effectively joined. For this to happen, we must acknowledge at the outset that manpower indeed possesses characteristics that make its contribution to strategy both irreplaceable and unique.

First, *manpower makes decisions*. Even with advances in data processing and artificial intelligence, strategic interaction promises to remain a largely heuristic enterprise requiring human response to the unexpected and unforeseeable. So long as the "fog of war" demands the sort of "creative stupidity" that machines seem yet so ill equipped to replicate, decisionmaking will remain manpower's most vital function.

Second, *manpower operates equipment*. Despite the prevalence of Buck Rogers-like visions of a fully automated battlefield saturated with vast arrays of remotely operated weapons, the gestation period for such technological advances is likely to be slow—perhaps excruciatingly so. Thus, for the foreseeable future, man and machine will remain inseparable partners.

Third, *manpower moves from place to place*—thus facilitating the sort of physical interaction and presence necessary to compensate for shortcomings in communications and control technologies.

Fourth, *manpower controls territory*—a necessary geostrategic imperative that seems destined to remain forever the province of humankind.

Fifth, *manpower occupies space* that otherwise could be occupied by something else. This underscores the existence of spatial occupancy limits, and it raises the question of whether available space is best occupied or left unoccupied in the face of particular threats and environmental conditions.

Sixth, closely related to the above, *manpower presents targets* that eventually may become casualties—the ultimate determinant of success or failure in conflict. This gives particular salience to the classical duality between concentration and dispersion. As forces compress to achieve mass and local superiority, they also present a larger, more vulnerable target. In contrast, as they expand to reduce vulnerability by creating a target servicing problem, they become commensurately less able to achieve local superiority.

Seventh, *manpower consumes resources.* As a living organism or system, manpower expends energy and thus must consume replenishing resources—food, dollars, ammunition, lodging, transportation—to remain viable. A dependency relationship, therefore, is created, the form of which exerts a decisive influence on the goals established for and by the system as well as on the manner in which it attempts to achieve those goals. Accordingly, this tends to be the major physical constraint on strategic interaction.

Lastly, and perhaps most importantly, *manpower is the ultimate manifestation of national commitment*—the forgotten social dimension of strategy to which British historian Michael Howard has attached so much significance.[2]

These atributes effectively establish the uniqueness of manpower and provide the underlying rationale for its incorporation into strategic planning. Yet, such factors have figured only remotely in the major manpower debates of our age. Manpower *qua power* has commanded far less attention as an object of analysis in recent years than has the family of questions more appropriately labeled personnel management or human resources development—the lingua franca of the business community.

We would do well to bear in mind that, although the etymological origins of the term *manpower* are notably obscure, the root word is clearly *power*, while *man-* is but a clarifying prefix. This not only establishes the terminological legitimacy of human capital as a major national asset, but also drives home the realization that to deal effectively with the strategic dimension of military manpower, we must more fully understand the nature of power itself.

THE NATURE OF POWER

It has been said that power is to politics what wealth is to economics. To the extent that strategy owes its legitimacy to the political ends for which it is pursued, then power logically is the quintessence of strategy. This relationship has been articulated with most cogency by retired Rear Admiral Henry Eccles, who defines strategy as "the comprehensive direction of *power* to *control* situations and areas in order to attain objectives [emphasis added]."[3]

The classic description of power as "deference value," put forth some thirty-five years ago by Harold Lasswell and Abraham Kaplan, continues to offer singular rudimentary appeal today. Simply stated, it suggests that to have power is "to be taken into account in others' acts (policies)."[4] But power is not merely possessed; nor is it merely acknowledged by those at whom it is directed. It is exercised, and, in the exercising, it induces particular behavior, including inaction. Therefore, in the interest of semantic (and conceptual) robustness, power shall be treated here as *the exercise of influence to control the behavior of another party and bring about obedience to, or compliance with, one's will.*

The defining characteristics of power are several. It may operate anywhere on the continuum of social interaction, from coercion involving fear and intimidation to persuasion involving respect and willing deference. Power depends for its success on leveraging the vulnerabilities and insecurities of a given adversary and exploiting his "capacity for belief." It need not be a conscious, purposive act. In fact, true power exists in its most unadulterated form independent of intentionality. Its existence will prompt others to take its likely effects into account in determining their own actions, even when its possessor does not expressly seek such deferential behavior. Nonetheless, the truly effective employment of power requires careful and thorough orchestration.

Conceptually, two factors distinguish interstate from interpersonal power: (1) physical proximity and (2) the stakes involved. Consequently, the lessons drawn from the study of power relations at one level may not always be applicable to power relations at the other level. Interstate relations obviously do not entail the face-to-face interaction characteristic of most interpersonal relationships. As distance between the parties to an encounter increases, so too do the barriers to accurate communication and, accordingly, the chances for misperception. (*Distance*, referring here to physical separation, also implicitly connotes psychic separation born of such factors as culture, language, and ideology.) On the other hand, distance also provides more latitude for manipulating the incongruence that exists between perception and reality. In other words, because distance makes objective reality less discernible and less measurable, perceptual factors come to play a more dominant role.

The effect this has on the stakes involved is somewhat convoluted. Although in an absolute sense the stakes associated with interstate relations are by far greater, the stakes at the interpersonal level are more direct, more tangible, and thus more real to the parties involved. Therefore, the application of power may have to be adjusted to accommodate differing conceptions of risk under different circumstances. Risk at the interpersonal level, being highly individualized and concentrated, is subject to visceral tendencies not so evident at the interstate level, where risk tends to be collectivized, diffused, and accordingly less calculable.

Perhaps because the stakes at the interstate level are demonstrably greater, even if less calculable, the most effective use of power at this level involves tacit, rather than explicit, threats. Such tacit threat-making, the *modus vivendi* of superpower relations in the modern era, is appealing because it avoids the costs and risks of having one's bluff called. Depending on the circumstances and the potential strategic payoffs, the appearance of either causation or denial can be created *a posteriori* with relative ease.

In pure theoretical terms, power is a relative, relational concept involving interaction between two particular parties and tailored to accommodate the attitudinal and behavioral propensities of those parties according to the peculiar circumstances of the moment. However, because of the extraordinarily large number and type of possible variables and interactions, real-world strategic planning must be based on the generalizability of power across a broad range of threats and contingencies. Thus deterrence, probably the most common form of contemporary power politics, has shown itself to be fundamentally indivisible across the entire spectrum of conflict; where deterrence is not operative, conflict naturally will seek that level for actualization. This phenomenon of indivisibility tends to affect and to be affected by one's conception of the future nature of conflict and the means considered appropriate for use in each case.

At the nuclear level, which is technology exclusive, manifest power has become virtually unusable, although latent power provides a permanent base of leverage supporting all other forms of strategic interaction. At the conventional level, where a semblance of balance exists between technology and manpower, technological advances have created such a blurring of the nuclear threshold that manifest power has become increasingly unusable in this domain as well. However, at the manpower-intensive subconventional level, the sophisticated use of power remains a singular challenge for coping effectively with what promises in the years ahead to be clearly the most prevalent form of conflict.

Considering the sometimes fuzzy-headed conceptualizations that have guided U.S. strategic planning in recent years, it is important at this point to distinguish between *power* and *force*—the former being at root a psychological phenomenon, the latter a purely physical phenomenon. Unfortunately, where total ignorance of this important distinction has not prevailed, there has been a tendency among practitioners to view the distinction as little more than academic pedantry. We would do well to recall, however, not only that this very distinction constituted the essence of B.H. Liddell Hart's conception of strategic indirection, but also that the classic articulation of the idea was provided well over 2,000 years ago by Chinese scholar Sun Tzu, who suggested: "Supreme excellence consists in breaking the enemy's resistance without fighting."[5]

Hans Morgenthau, of course, provided the most penetrating contemporary formulation of the concept when he observed:

> Political power must be distinguished from force in the sense of the actual exercise of physical violence.... When violence becomes an actuality, it signifies the abdication of political power in favor of military or pseudo-military power.... The actual exercise of physical violence substitutes for the psychological relation between two minds, which is of the essence of political power.[6]

Thus, power works on the psyche rather than on the physical senses. In the main, it relies on demonstration rather than actual engagement, although too literal an interpretation of Morgenthau's prescription is likely to produce a flawed instrument of persuasion. What actually is sought is the intentional, selective use of force under conditions of one's own choosing so as to obviate the unwanted necessity of having to use force under conditions not of one's own choosing.

For this conception to be appreciated fully, we must acknowledge that power has three major components that complement and in fact compensate for one another. The first of these components is *capability*, which manifests itself in both force structure (the operational instrumentality of manpower) and technology. The second component of power is *intentionality*, which manifests itself principally in doctrine (as well as in official policy pronouncements). Doctrine, though ostensibly prescriptive in function and intended for internal consumption, actually is likely to be construed by external audiences as *descriptive* of how one will fight. Thus, it has considerable propaganda value and can embellish one's image of strength and fortitude.

The third and most vital component of power is *will* or resolve, which manifests itself most assertively in *selective engagement*. In other words, there are acute limits to the resolve that can be demonstrated by such traditional measures as strident rhetoric and defense spending increases—the locus of most Reagan administration initiatives to date. Consequently, it periodically may be necessary actually to employ force in order to create tangible evidence of one's willingness to sacrifice. Needless to say, successful performance in such engagements is absolutely essential to the establishment and enhancement of one's reputation—which, as Klaus Knorr has suggested, is a critical determinant of one's putative military power.[7]

From these basic premises, we may offer two propositions:

- *Proposition 1:* Selective engagement, successfully executed, at lower levels of the conflict spectrum (e.g., a counterterrorist reprisal raid or a counterinsurgency campaign) can enhance one's position of power at higher levels of conflict, where engagement is infeasible or undesirable.

- *Proposition 2:* Demonstrated performance—successful or unsuccessful—by even a limited military force (such as a commando team) will be generalized by external observers as being representative of the parent state's overall fighting proficiency. In other words, there is a natural tendency to extrapolate from very limited data in making such assessments, especially where it reinforces existing preconceptions.

These three components of power—capability, intentionality, and will—underscore the symbiotic relationship that exists, or should exist, among force structure, technology, and doctrine. Technology, now and forever the central element of America's defense posture, has no "persona" in and of itself. Its character, whether passive or aggressive, is a function of both its users (manpower) and its intended uses (doctrine). Absent the human dimension, a given weapon system is nothing more than an inert assemblage of parts.

On the other hand, technology recognizably is a multiplier, designed to be used as a complement to, and an enhancer of, manpower. However, America's technological obsession and the country's collective attitudes regarding the sanctity of human life have engendered a belief that technology is a substitute for manpower and a panacea for most, if not all, of our military shortcomings. The most obvious recent example of this propensity is the dominant role accorded emerging technologies in the so-called Deep Strike and Follow-on Force Attack concepts now being instituted within NATO. Advanced target surveillance and tracking systems, long-range precision-guided delivery systems, and various guided and unguided submunitions are being developed as the principal means for striking high-value air bases, command posts, and second echelon formations deep in Warsaw Pact territory. In contrast, the Soviets have placed major emphasis on the use of human formations—Operational Maneuver Groups and special operations forces—for facilitating deep penetration strikes into NATO's rear. The perceptual differences, as a reflection of the respective fighting tendencies of the two sides, are striking.[8]

Technology and manpower, though not directly fungible, can compensate to some extent for each other. Increased weapons lethality commonly is sought as a more efficient means than manpower of producing mass. Largely unrecognized, though, is the utility of manpower in compensating for technology. One need only look at the People's Republic of China, with its mass army and relatively unsophisticated technology, to see the most visible example of this phenomenon. This suggests that where there are upper limits to the numbers of weapon systems that can be fielded (due, say, to exorbitant costs or declining budgets), manpower increases may ameliorate the problem by adding to one's relative calculus of strength.

In this regard, the United States has created a Gordian knot for itself. Rising costs and other systemic constraints, such as the constipation of the weapons acquisition process, tend to limit the number of weapons that can be fielded at any one time. By the same token, the volunteer nature of the American force effectively limits the manpower that can be fielded (by constraining either supply or demand, or both). The inevitable result is a U.S. force posture widely viewed as too small to meet all of its global commitments and certainly at a numerical disadvantage vis-à-vis the Soviet Union.[9] All in all, this suggests the need for a much more creative distribution scheme for both manpower and weapons than currently exists.

One thing that obscures manpower strength is the technology intensiveness of a given force and the attendant personnel-to-platform ratios of the weapons constituting the force. Fighter aircraft, for example, are very efficient in this regard, for they require few personnel to provide a significant amount of killing potential (even though they do, in fact, require large hidden support structures). Tanks, artillery pieces, and other crew-served weapons are relatively worse in this respect, typically requiring on the order of four to six personnel for operation. Naval vessels are by far the worst systems from the standpoint of manpower visibility or consumption, for they require extraordinarily large numbers of personnel to operate relatively few platforms. Thus, with naval vessels the value of numerical manpower strength is diminished appreciably, if not lost altogether. To appreciate the simple elegance of this frequently neglected calculus, consider the effect of reducing the technological complexity and size of a given weapon system. Say this particular weapon, which used to cost $9 million apiece and required a crew of 750, now, in a more streamlined variant, operates with a crew of 250 and costs only $3 million. Obviously, we can now field three simpler versions where previously we could field only one. Not only is our putative power increased threefold, but, generally speaking, we can increase our geostrategic coverage by deploying to three different locations simultaneously.

Doctrine also plays an absolutely crucial role in establishing the importance of manpower and its relationship to technology. More importantly, it is, especially in the absence of actual hostilities, the principal vehicle for creating the fighting image of a force. As Soviet doctrine has demonstrated so well, a forcefully stated doctrine that emphasizes the importance of such things as offensive action, surprise, and victory can go a long way toward creating the image of a soldier that appears to be ten feet tall. The West's greatest fears about the awesome Soviet military juggernaut derive not from the actual performance of that force but from its size and, more importantly, from our own readings of Soviet doctrine.[10] The opaqueness of Soviet society leaves outsiders little choice but to rely on surrogate sources of insight, such as doctrinal writings, for one's definition of reality.

The centrality of doctrine in the hierarchy of Soviet military thought tends to endow it with a degree of legitimacy that conveniently removes it from consideration by outsiders as propaganda, per se. Almost effortlessly, therefore, it becomes received truth. But when doctrinal pronouncements are accepted uncritically as a sort of holy writ offering irrefutable "proof" of the inner workings and hidden mechanisms of the Soviet military machine, then it is we ourselves, as the recipients of the information, who have created the reality of the situation. The Soviets, recognizing this, have exploited such tendencies to increase their power without ever firing a shot.

In contrast, the U.S. military establishment has shown itself intellectually ill equipped to come to grips with the internal contradiction of espousing a warfighting doctrine for external consumption while simultaneously appeasing domestic and allied constituencies who construe such pronouncements as blatant warmongering.[11]

The operative strategic principle, then, that should guide our thought and action in the contemporary era is to *espouse Clausewitz*—with his emphasis on warfighting and victory—but to *practice Sun Tzu*—with his emphasis on the application of extraordinary, or indirect, force.[12] Though seemingly paradoxical on the surface, such an approach embodies an underlying sophistication ideally suited to the demands of the media age in which we live. As such, it fully captures the essence of modern-day strategy.

Perceptions Management: The Essence of Strategy

In the final analysis, power—the quintessence of strategy—is an exercise in *perceptions management* (or, dare it be said, *consciousness engineering*): the manipulation of symbols to create desired images in the minds of particular target audiences. While moralizers and psychoanalysts alike may be inclined to question both the propriety and the feasibility of such pretense, it is fatuous to suggest that psychological preparation cannot be achieved through careful orchestration. Is this not, after all, the fundamental premise behind the use of propaganda? As Robert Jervis has opined: "If a policy is to have the desired impact on its target, it must be perceived as it is intended."[13]

Kenneth Boulding tells us that decisionmakers respond not to the *objective* facts of a situation, whatever that may mean, but to their *image* of the situation. The image is the total cognitive, affective, and evaluative structure of a behavior unit, or its internal view of itself and its universe.[14]

The symbol, in turn, as described by Murray Edelman in his important book, *The Symbolic Uses of Politics*, stands for something other than itself, and it also evokes an attitude, a set of impressions, or a pattern of events associated through time, space, logic, or imagination with the symbol. A symbol's meaning is not inherent in the symbol itself but rather in observers and their social situations. Symbols elicit, in concentrated form, those particular meanings and emotions that the members of a group create and reinforce in each other.[15]

This calls to mind Heinrich Hertz's Law, which states: "The consequences of the images will be the images of the consequences."[16] The effects of events depend in large measure on the picture we have conjured up in our minds beforehand of the significance of such effects. To appreciate the wisdom of this proposition, we need only consider the commonly held view of the devastating consequences of nuclear war—especially under a so-called nuclear winter scenario—and the effect that this image has had on our attitudes concerning the utility of nuclear weapons.

Erving Goffman has described one of the five basic moves of human expression games as "the control move": the intentional effort of an informant to produce expressions that he thinks will improve his situation if they are gleaned by the observer. In other words, a given individual, aware that his actions, expressions, and words will provide information to an observer, will attempt to create an advantageous definition of the situation. He does this by assuming the viewpoint of the observer in order to discern how the observer will respond. Goffman calls this "impression management" (or what here has been termed *perceptions management*). What is involved essentially is not communication but rather a set of tricky ways of sympathetically taking the other into consideration as someone who assesses the environment and might profitably be led into a wrong assessment (or a right one despite his suspicions, ignorance, or incompetence). The various processes of control do not strike at the observer's capacity to receive messages, but at something more general, his ability to read expressions. Thus, when the subject employs verbal means to convey information about his intended course of action, the observer—if he is properly to judge the significance of these communications—will have to attend to the expressive aspects of the transmission as a check on semantic content. Similarly, in trying to conceal while communicating, the subject, too, will have to attend to his own expressive behavior. A message, then, suggests Goffman, functions merely as one further aspect of the situation that must be examined carefully and controlled carefully in this contest of assessment between subject and observer.[17]

The conception of power put forth here owes its theoretical heritage not to the realist school, which has spawned most proponents of power politics to date, but to symbolic interactionism and phenomenology. Key among the defining characteristics of this philosophical orientation are the following precepts:

- Reality is socially constructed in the mind of man. It does not exist in any objective sense.
- Situations are defined and given meaning by humans. The definition of a situation, as distinct from the existential situation, is the operative factor in how persons construct their behavior.
- Symbolic communication is the basis of all social action. It is a two-way process with both parties giving each symbol meaning.
- Action in progress is interpreted by capturing the meanings persons attach to their conduct, taking the role of the other in gaining an objective view of the self, and constructing action as an intelligent response.
- Meanings are closely associated with collective experience among those using the symbols. Shared experience produces common insight.[18]

The basic assumptions underlying this school of thought, particularly its symbolic interactionist component, are perhaps best summarized in the work of Arnold Rose. First, notes Rose, "man lives in a symbolic environment as well as a physical environment and can be 'stimulated' to act by symbols as well as by physical stimuli." Second, "through symbols, man has the capacity to stimulate others in ways other than those in which he is himself stimulated." Third, "through communication of symbols, man can learn huge numbers of meanings and values—and hence ways of acting—from other men."[19]

Power assessment, the largely intuitive process by which one's leverage over others is determined and conferred, operates on the basis of a *lowest common denominator principle:* What is important is what is most understandable to the widest audience. Thus, what count are tangible factors that are amenable to quantification and observation: numbers, size, composition, disposition, and demonstrated performance. Intangible qualitative factors—such as leadership, training, morale, discipline, cohesion, and individual technical aptitudes—though intuitively important and frequently invoked by pseudo-sophisticates claiming special insight into the dynamics of conflict, nonetheless are not part of the normal calculus of power. They come into play only when force actually is employed, and even then the precise nature of their respective contributions to particular outcomes is extremely difficult to gauge. Accordingly, this leads us to yet another proposition:

- *Proposition 3:* Intangible qualitative factors are assumed by external observers to exist in the absence of demonstrated evidence to the contrary.

It is precisely for this reason that, despite the perceptive observations of some analysts concerning the leadership, morale, competence, and drug and alcohol abuse problems that afflict the Soviet army, we continue to imbue our principal adversary with seemingly superhuman fighting qualities. It also is for this reason that, on those selected occasions when force is employed, the success of the user is absolutely imperative.

If we are to adopt a conception of power that adequately recognizes the importance of symbolism and imagery, and to exploit the human dimension of this conception to the fullest, we must acknowledge the nature of the environment in which we live and operate. Marshall McLuhan and Quentin Fiore, in expressing their vision of war and peace in the "global village," have averred: "The electronic culture of the global village confronts us with a situation in which entire societies intercommunicate by a sort of 'macroscopic gesticulation'."[20]

The policy imperative that presents itself is to concentrate first on reinforcing existing images rather than attempting to create new ones or to alter strongly held old ones. Boulding has pointed out that the image is built up as a result of the past experiences of its possessor. The meaning of a message, therefore, is the change it produces in the image. But images are inherently resistant to change. When we receive messages that conflict with an established image, our initial impulse is to ignore the messages altogether or to reject them as untrue. The subjective knowledge structure possessed by an individual or group has both factual and value-based dimensions. If a message is received that is neither good nor bad, it may have little or no effect on the prevailing image. If it is perceived as bad or hostile to the prevailing image, there will be resistance to accepting it. On the other hand, messages that are favorable to the existing image are received easily, and even though they may cause minor modifications of the knowledge structure, there will not be any fundamental reorganization.[21]

Boulding's views on the nature and durability of images are supported by an exceptionally large body of communications research and theory. Collectively, these findings demonstrate that if we can identify and define the images held by a particular target group, we can influence the attitudes and beliefs of that group by playing on those images. In the present context, we must begin by acknowledging that, contrary to the Clausewitzian dictum that "war admittedly has its own grammar, but not its own logic," in point of fact the military does have both its own grammar *and* its own logic. It is a more or less universal logic, with a supporting value structure, that cuts across those nationalistic, cultural, and linguistic lines that otherwise are the enduring impediments to international communication. Again, to cite McLuhan and Fiore:

> Games are fashions . . . because they involve the sensory life of a society in a mocking and fictitious way. To simulate one situation by means of another one, to turn the whole working environment into a small model, is a means of perception and control by means of public ritual.[22]

The game at hand here is the Game of Machismo—a universalistic, traditional conception of war and man's role in war. It may well represent the next stage in the evolution of human aggression, in which the thrill of the hunt displaces the actual capture as the ultimate form of psychic gratification.

The Game of Machismo operates at two levels: the individual level and the collective level. At the individual level, the normative image is the classical warrior ideal—the courageous fighter, inured to hardship, willing to sacrifice, and equipped with such martial virtues as soldierly bearing, loyalty, and obedience to authority. In the modern era, the model for this image is far less likely to be represented by the chivalric Lancelot type than by a place on what somewhat facetiously may be called the "Ninja–Hell's Angels Continuum." Thus, the archetypal image may range from the stealthful assassin type—swathed in black and possessed of the most deadly arts, who stalks his prey in terror-inducing silence—to the swaggering, tattooed-bully type, garbed in black leather and metal studs, whose confrontational, no-nonsense style is embodied in the challenge, "Go ahead, punk, make my day."

At the collective level, the normative image, applied to the fighting force as a whole rather than to the individual warriors constituting the force, is an aura of strength that connotes confidence, competence, and resolve. The desired image at this level should demonstrate three distinctive but interrelated attributes:

- Fighting spirit, the principal vehicle for which is doctrine
- Fighting capacity, which is a function of force structure and technology
- Fighting prowess, which is demonstrated on those occasions when force actually is employed

To review the bidding at this point, we have sought to establish how symbols related to the human dimension of military power can be manipulated more effectively to reinforce widely held images of soldierly acumen and thereby enhance perceptions abroad of U.S. strength. In the most fundamental sense, this suggests an assiduous effort to exploit extant myth and ritual—an enterprise in which we certainly can learn something of value from purportedly primitive societies that have successfully displaced war with ritualistic behavior. Such cultures, recognizing the prohibitive costs of physical aggression, have created elaborate rituals involving dress, appearance, and gesticulation to provide outlets for their aggressive drives and to establish winners and losers, all within reasonable, controllable bounds of social conduct.[23]

The two areas most conducive to simple symbol manipulation are military dress and ceremony. McLuhan and Fiore have discussed at some length the idea of clothing as weaponry, designed to combat hostile conditions by harnessing human energy.[24] This is more than metaphorical gibberish, for all cultures and subcultures use clothing to signal something about themselves to others. The military, after so many years of practice, unquestionably should be the most sophisticated institution of any in this regard. Unfortunately, ours is not. For too long, we have sought functionalism and cost effectiveness in the design and procurement of uniforms, without fully recognizing the power potential of such things as camouflage or tiger-stripe battledress, distinctive headgear (berets, bush hats, forage caps, etc.), and even footwear (e.g., jackboots).[25] Whatever the functional utility or disutility of such items, or their effect on morale, it is their demonstration value that warrants our more serious attention.

Much the same may be said of our functional approach to ceremonies. We have not duly acknowledged (as others seemingly have) how simple formation marching, particularly where the goose step is employed, and battle drill can feed an image of discipline and efficiency. That famous passage from Jean Lartèguy's popular novel about the French *paras* in Indochina and Algeria, *The Centurions*, comes immediately to mind. Colonel Raspèguy, about to assume command of the 10th Colonial Parachute Regiment in Algeria, observes:

> I'd like France to have two armies: one for display, with lovely guns, tanks, little soldiers, fanfares, staffs, distinguished and doddering generals, and dear little regimental officers who would be deeply concerned over their general's bowel movements or their colonel's piles: an army that would be shown for a modest fee on every fairground in the country. The other would be the real one, composed entirely of young enthusiasts in camouflage battledress, who would not be put on display but from whom impossible efforts would be demanded and to whom all sorts of tricks would be taught. That's the army in which I should like to fight.[26]

In the abstract, there is little in this statement with which military idealists would quarrel. In fact, the natural tendency is to say, "Right on! Tell it like it is!" But, in this modern age of warfare by deterrence, Raspèguy's views are not totally in step with the unique demands of the time. Public perceptions are molded on the basis of scraps of information that to the purist or expert may bear little relevance to what really counts. For example, knowledgeable observers are quick to discern that European attitudes concerning the quality of U.S. troops in NATO are based not on the military proficiency of those troops but on the impressions they leave as members of the community. Because domestic and foreign publics rarely see our fighting formations in action, the image projected by soldiers on parade is the one likely to take root. There should be little doubt that the image we want projected is that of the experienced warrior—the hardened look, suntanned with muscles abulge, dressed in camouflage—rather than the pimply-faced raw recruit with oversized collar and plastic-looking shoes. Thus, even if we could afford two armies, it is the fighting army that we would want to do the ceremonial marching as well.

Selected Manpower Issues: Gauging Their Perceptual Impact

Beyond the fundamental considerations enumerated above, it is important, if we truly are to come to grips with the strategic dimension of military manpower, to be better attuned to the perceptual ramifications of various force structure alternatives. Ideally speaking, force structures should be designed to support established objectives, doctrine, and state-of-the-art technology in the face of identified threats and environmental conditions. Viewed in this sense, force structure is little more than an emergent, subsidiary activity that derives from larger strategic imperatives. The reality, however, deviates considerably from the ideal; it is a process constrained by various resource limitations, by competing political interests, and by bias, parochialism, and general bureaucratic inertia. The upshot of this is that force structure affects the governing environment every bit as much as it is affected by it. Our key to the future, therefore, and our ability to assert ourselves strategically, may well lie in our capacity for putting forces in place that are designed and manned with perceptual—as well as with economic, political, and bureaucratic—considerations in mind. An abbreviated survey of pertinent issues may serve to bring the challenge facing us more clearly into focus.

Light versus Heavy Forces. Whether forces should be essentially light or heavy in design is a question that looms ominously over the contemporary strategic landscape. The fundamental juxtaposition, of course, is between forces that are (and are seen to be) flexible, mobile, and easily projected to distant areas—light forces—and those that not only have the ability to engage in slugging matches but also are endowed with staying power—heavy forces. Despite recent flirtations with light infantry divisions, the United States over time has shown itself to have a clear preference for heavy forces (witness the navy's efforts under the Reagan administration to refit its fleet of battleships). Our fixation with conventional conflict in Europe, our associated penchant for preparing for past wars, and the inexorable march of technology all have fed this bias. However, barring unforeseen and highly unlikely developments in Europe, the wave of the future would seem to favor lighter forces that are suited for preemptive power projection—the ability to project a credible combat contingent into a disputed area with sufficient speed and surprise to present a prospective opponent with a *fait accompli*.[27]

Tooth-to-Tail Ratios. Although the historical record offers convincing evidence that our army has been hampered in the initial stages of past wars by inadequate service support, peacetime deterrent requirements underscore the value of emphasizing combat (tooth) over support (tail) capabilities. Generally speaking, a high tooth-to-tail ratio presents an image of a hard-hitting, mobile force with great freedom of action and relatively low targeting vulnerability. This visage of a lean, mean, streamlined force prepared to endure hardship and sacrifice reinforces the warrior image. On the other hand, the force is likely to be perceived as having little sustainability. In contrast, a force with a low tooth-to-tail ratio, though having greater sustainability, may appear sluggish and inflexible. Ultimately, of course, the reality will depend in large measure on doctrine. The Soviets, for example, though having a much higher tooth-to-tail ratio than U.S. combat units, depend on a logistical system fully compatible with their use of echelons or waves that effectively pushes support from rear to front during the course of combat operations.[28]

Active versus Reserve Forces. Reliance on a large standing force obviously serves the cause of deterrence well, because such a force is much more likely to be perceived as stronger in an immediate sense, better able to influence short-fuse political situations, and more flexible. Regardless of the reality of the situation, it is natural to expect the active component to be better trained, equipped, and maintained as well as more responsive than a force dependent on mobilizable reserves. On the other hand, the former also will be seen as less sustainable over the long haul—although sustainability becomes essentially moot if one assumes that prolonged combat is unlikely to occur. Almost universally, reserves are perceived as second-rate (or even third-rate) fighting forces—undermanned, ill equipped, ill trained, and undisciplined; civilians merely masquerading as soldiers. Only in those instances where a bona fide mobilization doctrine exists to legitimize the use of reserves, and where those forces periodically are activated, is this perception likely to be altered. The Swiss and the Israelis are cases in point. Absent these two preconditions, heavy reliance on reserves—such as that practiced by the United States—is highly unlikely to be credible to prospective adversaries.

Elite versus Regular Forces. Traditionally, elite forces have been distinguishable from regular forces by such factors as voluntarism, special selection criteria and training, and distinctive clothing or insignia. They have tended to be misused by higher level commanders and policymakers, and critics have viewed them as a drain on the leadership and high-quality manpower of regular units.[29] Nonetheless, it commonly is recognized that elite forces maintain high levels of readiness, display initiative and aggressiveness, are extremely well trained for the execution of highly specialized and difficult assignments, and are emotionally resilient, courageous, and physically fit. Furthermore, they are seen as totally committed and more willing than regulars to enter into combat, regardless of the nature of the contingency. Thus, even though their usual light configuration equips them poorly for sustained or heavy combat, they provide an exceptional deterrent capability. They are ideal for preemptive power projection and, against appropriately limited objectives, they greatly increase the chances of successful outcomes so essential to effective selective engagement.

Voluntarism versus Conscription. Under current all-volunteer conditions, this issue is seemingly academic. Nonetheless, the future demands that we consider the perceptual ramifications of the issue fully. If the fulfillment of total manning requirements could be assured, an all-volunteer force would seem to provide the best deterrent effect. The image created by a volunteer force is that of a body of highly motivated individuals dedicated to the national purpose and willing to sacrifice. Failure to attain prescribed manning levels, however, exposes to full view the overriding negative effect of voluntarism: that the pool of available manpower actually is limited and thus may hamper the overall ability of the force to engage in and sustain combat. Interestingly enough, though, conscription is sufficiently widespread and historically based that it carries with it no particular stigma. If anything, it is a measure of state control over the masses. The advantage of a conscript force is that it presents the image of a virtually inexhaustible manpower supply that can be tapped at will to bolster one's strength. Furthermore, it may counteract a potentially problematical aspect of the all-volunteer concept: that being, in the opinion of at least one observer, that such voluntarism by the United States may inadvertently signal to others—particularly the Soviets—that "America has reached the next-to-last step in decadence with a quasi–mercenary army and [with] no intention to fight another war."[30] So far, that fear does not seem to have been borne out, but only time will tell.

Female Content. In recent years, the number of women in the U.S. military has increased markedly—a trend that seems likely to continue in the years immediately ahead. Although in actuality increased numbers of women will free more males for combat duties, the image abroad of the sexual integration of U.S. forces may have unintended consequences. Traditionalist views of male supremacy remain widespread in virtually all societies. The time is yet to come when women will be fully accepted as other than the weaker sex, whose only legitimate function is procreation. It has been suggested, for example, that the Soviet Union might be inclined to view an expanded military role for women in the United States as a sign of weakness brought on by recruiting difficulties. The Soviets, it is argued, would be unlikely to appreciate the equal rights rationale behind such integration.[31] We ourselves have subscribed to this view in our own assessments of increased female content in the Soviet armed forces. Other cynical voices have been heard on the subject. One Israeli military officer, for example, has been quoted as saying: "We don't put women in front lines because we don't want them killed. You don't have a draft or registration and you're debating whether to send women into combat. . . . You spend more time arguing about women than trying to mobilize the best men."[32] Whatever contribution women might actually make to the effectiveness of U.S. forces, their increased role raises thought-provoking questions concerning how their utilization will be perceived by others.

Racial/Ethnic Composition. The racial and ethnic composition of the American military raises similar concerns. The fact that a significant percentage of the total U.S. force is made up of minorities has been the object of considerable scrutiny during the all-volunteer era—though not necessarily for *strategic* reasons. Three factors in particular converge to make this issue significant perceptually. First, past racial fractionation within the U.S. armed forces may be perceived by outside observers as still prevalent, though perhaps submerged. Second, it is not unreasonable to surmise that others, most notably the Soviets, are especially sensitive to the potentially destabilizing and disruptive effects of the issue because of ethnic divisions within their own armed forces and societies. Third, other cultures in which racial and class stratification is more commonly accepted may question the ability of minority personnel to operate the sophisticated equipment that dominates today's battlefield. Even our own NATO allies—particularly the Germans—reportedly have expressed concern about the number of blacks in our Europe-deployed units and have complained of problems stemming from "cultural differences."[33] Yet another strategic concern, not principally perceptual in nature, to which we must be sensitive is the potential effect that latent ethnic and racial loyalties might have on U.S. interventionist tendencies. For example, would the fact that we have a sizable Hispanic contingent in our armed forces affect our ability to intervene, if necessary, in Latin America?

Unionization. Since November of 1978, when the prohibition of military unions in the United States was signed into law, unionization has been essentially a dead issue. It is likely to remain so in the immediate future. Nevertheless, the domestic political forces that originally brought the issue to the fore are unlikely to remain dormant forever. As peacetime conditions continue, and as the military services are forced to compete more vigorously with the private sector for available manpower, the demands for occupational conditions and privileges comparable to those provided elsewhere will increase. It is almost inevitable, therefore, that further calls for unionization will be heard. While military unionization will be viewed by civil libertarians as singularly progressive in nature, it is difficult to dispel the notion that it will be viewed by others, particularly by authoritarian regimes abroad, as anything other than a source of internal disintegration and fragmentation. To the Soviets, whose concerns with combat morale, political indoctrination, troop control, and the proper relationship between labor and management in general are well documented, a move to unionization by the United States almost certainly would be construed as representing a breakdown in cohesion and effectiveness. Accordingly, deterrence seemingly would not be well served by such a move.

Conclusion

In the final analysis, the strategic dimension of military manpower involves the effective management of perceptions—principally external perceptions. We must recognize, however, the inherent sensitivity of issues such as those discussed above, and acknowledge that the logic put forth here may well fly in the face of domestic social imperatives. What seems strategically prudent with regard to female and racial content, elitism, and like issues may bear little logical relation to what is right and feasible domestically. Furthermore, we should be sensitive to the fact that if we are too obvious about our manipulation of symbols, our words and actions may appear cosmetic and contrived, thereby totally losing their value. This has been a major failing of the Reagan administration, which, though more sensitive than most recent administrations to the importance of political and military symbolism, nonetheless on numerous occasions has erred by being blatantly obvious in its intent.

Given these cautions, we also should be more attuned to the nature of the environment in which we live—especially insofar as the suggestion is concerned that, by orchestrating the instruments of power at our disposal, we can avoid having to use force when it is not to our advantage to do so. It is a characteristic of American society that our public ethos and the positivist scientific ideal—to which we tend, on the whole, to subscribe—are remarkably congruent. We are seekers of truth. It is time we thought more seriously about becoming shapers of truth.

NOTES

1. Edward N. Luttwak, "The American Style of Warfare and the Military Balance," *Survival* (March/April 1979): 57-60.
2. Michael Howard, "The Forgotten Dimension of Strategy," *Foreign Affairs* (Summer 1979): 975-86.
3. Henry E. Eccles, *Military Concepts and Philosophy* (New Brunswick, N.J.: Rutgers University Press, 1965), p. 18.

4. Harold D. Lasswell and Abraham Kaplan, *Power and Society* (New Haven, Conn.: Yale University Press, 1950), pp. 74-102.
5. Sun Tzu, *The Art of War*, trans. Samuel B. Griffith (London: Oxford University Press, 1963), p. 77. Also see B.H. Liddell Hart, *Strategy* 2d ed., rev. (New York: Praeger, 1967).
6. Hans J. Morgenthau, *Politics Among Nations* 5th ed., rev. (New York: Alfred A. Knopf, 1978), p. 31.
7. Klaus Knorr, *Military Power and Potential* (Lexington, Mass.: D.C. Heath, 1970), pp. 3-8.
8. See Daniel Gouré and Jeffrey R. Cooper, "Conventional Deep Strike: A Critical Look," *Comparative Strategy* 4, no. 3 (1984): 215-48; and Boyd D. Sutton et al., "Deep Attack Concepts and the Defence of Central Europe," *Survival* (March/April 1984): 50-70.
9. See Jeffrey Record, *Revising U.S. Military Strategy: Tailoring Means to Ends* (Washington, D.C.: Pergamon-Brassey's, 1984).
10. The number of sources that represent little more than regurgitations of Soviet doctrinal literature is too vast to list here. For critical assessments that take a contrary point of view and point out internal problems of the Soviet armed forces, see Andrew Cockburn, *The Threat Inside the Soviet Military Machine* (New York: Random House, 1983); and Richard A. Gabriel, *The New Red Legions* (Westport, Conn.: Greenwood Press, 1980).
11. For a cogent treatment of the efficacy of a war-fighting doctrine, see Colin S. Gray, "War-Fighting for Deterrence," *The Journal of Strategic Studies* (March 1984): 5-54.
12. For a useful comparison of the precepts of Clausewitz and Sun Tzu, see Tomas Ries, "Sun Tzu and Soviet Strategy," *International Defense Review* 4/1984): 389-92.
13. Robert Jervis, "Deterrence and Misperception," *International Security* (Winter 1982/83): 3-30.
14. Kenneth E. Boulding, "National Images and International Systems," *Journal of Conflict Resolution* (June 1959): 120-31.
15. Murray Edelman, *The Symbolic Uses of Politics* 2d ed. (Urbana, Ill.: University of Illinois Press, 1985), pp. 6-14.
16. Marshall McLuhan and Quentin Fiore, *War and Peace in the Global Village* (New York: Bantam Books, 1968), p. 16.
17. Erving Goffman, *Strategic Interaction* (Philadelphia: University of Pennsylvania Press, 1969), pp. 12-17.
18. Among the most penetrating treatments of the rudiments of symbolic interactionism are George Herbert Mead, *Mind, Self and Society*, ed. Charles Morris (Chicago: University of Chicago Press, 1934); George Herbert Mead, *The Philosophy of the Act*, ed. Charles Morris (Chicago: University of Chicago Press, 1938); and Herbert Blumer, *Symbolic Interactionism: Perspective and Method* (Englewood Cliffs, N.J.: Prentice-Hall, 1969). The basic precepts of phenomenology derive principally from the

works of Edmund Husserl and Alfred Schutz. See Edmund Husserl, *Phenomenology and the Crisis of Western Philosophy* (New York: Harper & Row, 1965); and Alfred Schutz, *Collected Papers*, 3 vols. (The Hague, Netherlands: Martinus Nijhoff, 1971).

19. Arnold Rose, "A Systematic Summary of Symbolic Interaction Theory," in *Human Behavior and Social Processes*, ed. Arnold Rose (Boston: Houghton Mifflin, 1962), pp. 5–9.
20. McLuhan and Fiore, *War and Peace*, p. 17.
21. Kenneth E. Boulding, *The Image* (Ann Arbor: University of Michigan Press, 1956), pp. 3–18.
22. McLuhan and Fiore, *War and Peace*, pp. 168–69.
23. For an excellent and comprehensive study of ritualistic behavior associated with human aggression, see Irenäus Eibl-Eibesfeldt, *The Biology of Peace and War* (New York: Viking Press, 1979).
24. McLuhan and Fiore, *War and Peace*, pp. 157–68.
25. For a discussion of functionalism in American military dress, see Sue Mansfield, *The Gestalts of War* (New York: Dial Press, 1982), pp. 158–60.
26. Jean Lartèguy, *The Centurions* (New York: Avon Books, 1961), p. 266.
27. See Kenneth Allard, "Soviet Airborne Forces and Preemptive Power Projection," *Parameters* 10, no. 4 (1980), pp. 42–51.
28. See Steven L. Canby, "The U.S. Defense Policy: The Problem is not More Money," *AEI Foreign Policy and Defense Review* 1, no. 3 (1979), pp. 23–36.
29. For full discussion of the advantages, pitfalls, uses, and misuses of elite forces, see Roger A. Beaumont, *Military Elites* (Indianapolis, Ind.: Bobbs-Merrill, 1974); Roger A. Beaumont, "Military Elite Forces: Surrogate War, Terrorism, and the New Battlefield," *Parameters* 9, no. 1 (1979): 17–29; and Eliot A. Cohen, *Commandos and Politicians: Elite Military Units in Modern Democracies* (Cambridge, Mass.: Harvard University Center for International Affairs, 1978).
30. William F. Long, Jr., "Counterinsurgency: Corrupting Concept," *U.S. Naval Institute Proceedings* (April 1979): 57–64.
31. Martin Binkin and Shirley J. Bach, *Women and the Military* (Washington, D.C.: The Brookings Institution, 1977), p. 97.
32. Seth Cropsey, "Women in Combat?" *The Public Interest* (Fall 1980): 58–73.
33. William Raspberry, "Black Soldiers and the Defense of Europe," *Washington Post*, 16 June 1982, p. A15.

3 THE ROLE OF MANPOWER IN TRADITIONAL STRATEGIC THOUGHT

John Keegan

It is said that "God is always on the side of the heaviest battalions." This aphorism, one of the best-known and most-quoted pieces of conventional military wisdom, is usually but fallaciously ascribed to Napoleon. In fact, it comes from the pen of Voltaire, who lived in an era of warfare that military historians generally describe as a contest between regular armies, in which quality rather than quantity was the decisive factor.

But if Voltaire, a keen observer of the contemporary scene, who had observed the Seven Years' War, the War of Austrian Succession, and the American Revolution at close hand, detected that numbers had begun to count in the wars of his era, we may deduce either that the factor of quality had begun to decline in relative importance during the eighteenth century or that modern military historians have overestimated its significance in retrospect. It is certainly possible to argue the first of these propositions. Size of population greatly impressed the economists and social scientists of eighteenth-century France, as evidenced in their analyses of the world collected in Diderot's *Encyclopedia*. As citizens (not yet *citoyens*) of the richest, strongest, and most populous state in contemporary Europe, they unquestionably associated power with population. Bourbon France was an expansionist power, able to expand its boundaries not simply because it had good generals, productive arsenals, and a full treasury but because its reservoirs of manpower supplied the numbers needed to fill its army's ranks.

Historiographers would probably also concede the validity of the second proposition. Quality undoubtedly counted for a great deal in eighteenth-century wars. Frederick the Great's successes, in particular, can be made to look almost exclusively a function of Prussian military elitism.[1] But closer examination calls this proposition into question. For if Frederick was often victorious, he was also often defeated—defeated not by the failure of his powers of military command but by a preponderance of numbers. The battles of Rossbach and Leuthen were brilliant exercises in the art of maneuver against a stronger enemy. But at Hochkirch and Kunersdorf, Frederick was soundly beaten because the enemy outnumbered him, and he failed in his efforts to divide their forces.[2] In short, weaker armies could survive on an eighteenth-century battlefield only if they could avoid meeting the enemy toe to toe; thus Frederick's preference for the surprise attack and the maneuver in oblique order.

Frederick's success encouraged him, however, to mistake the phenomenon he observed for a general principle: that discipline was the prerogative of small powers and would always negate the numerical superiority of larger agglomerations, which by their nature could not achieve the same level of public efficiency. "I perceive," he wrote in his political testament, "that small states can maintain themselves against the greatest monarchies [i.e., France, Austria and Russia], when these states put industry and a great deal of order into their affairs. . . . [G]reat empires are full of abuses and confusion."[3]

In other words, militarily speaking, small is beautiful. And it was a truth that experience from beyond the cockpit of Europe seemed to bear out. The success of the British in India, where tiny armies of closely disciplined troops overcame armies too large to be counted, seemed to give credence to the idea. No less telling had been the testimony of Antiquity, when the militias of Greek warriors, notably in their victory over the Persians at Marathon, demonstrated that numbers did not count when the enemy was an unwilling and unfree subject of his commander.

The flaw in Frederick's analysis of the number phenomenon was his assumption that large states could not make themselves efficient. He dismissed the possibility because he thought the courts of large states incurably corrupt and so inevitably given to conspicuous consumption of the state's revenues. He did not, in short, anticipate the abolition of courts through revolution. Nevertheless, when revolution broke out in France in 1789 it generated one of history's greatest maximizations of military potential. After the shattering and unexpected overthrow of Prussia by Napoleon in 1806, Gneisenau wrote: "One cause above all has raised France to this pinnacle of greatness, the revolution awakened all her powers and gave to every individual a suitable field for his activity. What infinite aptitudes slumber undeveloped in the bosom of a nation!"[4]

Carnot's famous exhortation to the people of France at the outbreak of the Revolutionary Wars is familiar to all. Its effectiveness must be kept in perspective, as also must that of the system of conscription that flowed from it. Frenchmen were often as keen to avoid the draft as they were to ask what they could do for their country. But the size of the French army did increase astonishingly between 1789 and 1796, from about 156,000 in the last days of the ancien régime to about 400,000 on the eve of Napoleon's whirlwind irruption into Italy.[5]

The victories of 1796 were won by Napoleon's genius rather than his command of numbers. Thereafter his appetite for manpower grew apace with his ambitions. The demands of the conscription system on the French population became increasingly exorbitant; moreover, a distinctive feature of Napoleon's pattern of conquest was incorporating the armies of governments he had defeated into the French order of battle. Italian, Spanish, Dutch, and eventually a constellation of German contingents, including after 1807 the rump of the Prussian army, were made subordinate to the French high command and marched about Europe at its behest, eventually as far as Moscow.[6]

Despite the enormous size of the armies assembled by Napoleon, Clausewitz, the most famous interpreter of his strategic methods, did not lay critical emphasis on manpower as a key to Napoleon's victories. Without doubt, one of the best known of Clausewitz's dicta is that "the best strategy is always to be *very strong*; first in general and then at the decisive point." But the assumption underlying Clausewitz's whole treatment of war is essentially the eighteenth-century view that manpower is an inelastic commodity, the supply of which cannot easily or quickly be increased. Hence, Clausewitz attaches importance to the same elements of strategy exercised by Frederick the Great: maneuver to divide the enemy's forces, surprise, and whirlwind offensive action.

At the time that Clausewitz was writing, however, social, political, and administrative changes were in progress throughout western Europe that would dramatically increase the elasticity of manpower supply. Improvements in public health would enlarge substantially the stock of available manpower, the institution of effective census would increase the efficiency of governmental bureaucracies, and a steady movement towards universal education would enhance the state's ability to find and enlist the men it needed to fill the army's ranks. In addition, the idea of the citizen army had survived the French Revolution and begun to mingle with and reinforce the tide of nationalist feeling that would manifest itself in Germany and Italy in the mid-eighteenth century.

The French governments of 1815 to 1870, which had discarded the principle of universal service along with other dangerous ideas spawned by the men of 1789,[7] failed to recognize that future enemies were taking steps during the 1840s and 1850s to create mass armies. First Prussia, then the states that fell under Prussian influence, introduced terms of short-service conscription that subjected all fit men to military training, then later retained them as mobilizable reserves.

The result was stupefying. In 1866 a short-service Prussian army overwhelmed the traditionally long-service Hapsburg army in six weeks.[8] Observers minimized the importance of mass—as they did all supporting evidence from the American Civil War—by reference to the Prussian generals' superior disposition of their forces. By 1870, the issue could be avoided no longer. The French army, byword for professionalism and inheritor of the Napoleonic tradition, was overwhelmed on its own territory in seven weeks by a German army that made up for its lack of tactical skills by its vastly superior reservoir of soldiers. The era of the mass army had arrived.

It was no accident that Prussia launched this new era. For it was Prussia that had invented the system of short-service conscription and Prussia that detected the contradiction between autocracy and the citizen militia system,[9] both crucial to the development of effective mass armies. Let us, for purposes of explanation, diverge for a moment from the consideration of strategic thought to the typology of armed forces.

In broad terms, there are six different means for raising an army. (Let us disregard the reciprocal effect that military organization has on social systems, since that would require for exploration a separate paper.) The first method we may call the warrior principle, in which military functions are exercised by a particular class or caste; it may be a large class, as was the case in Zulu and Mongol societies, or a small class, as in feudal western Europe. The advantage of the warrior principle is that it breeds high levels of battlefield ferocity. Its disadvantages include the servile subordination of the non-warrior segment of the population and a marked inability to adapt to technical change. The second method, characteristic of theocratic societies where the ruling class nurtures taboos about the shedding of blood particularly between fellow believers, is the slave army.[10] The slave army must contend with the contradiction that nominal slaves may inevitably come to exercise great power, often in an arbitrary and self-interested manner. But, as the history of the Ottoman and Mameluke states demonstrated, slave armies can produce a high level of both state authority and military prowess. The third method is the mercenary system, which promises a great deal in the short term but has obvious long-term disadvantages. The fourth is the regular principle, which we will describe as the institutionalization of mercenary service. The regular principle produces the most effective and docile of all armies, but it yields trained reserves in a trickle rather than a flood, since the regular's career interest is in prolonging his term of service for as long as possible. The fifth method is the militia system, which produces the opposite effect of the regular principle. Because all male members of the political class are liable to serve, available manpower is limited only by factors of population size and age distribution. But because the soldiers are citizens, whose productive lives are led outside the ranks, the voting tendency is toward attenuation of active service and so toward low standards of training. The sixth and last method is the conscriptive system, which may be regarded as a form of state taxation on the male population's time rather than money. Like all taxes, conscription is unpopular and may or may not yield a fruitful return on the effort spent collecting it.

Until 1857, Prussia had run a mixed conscriptive-militia system (mixed systems, it is worth noting, are quite common). The 1857 accession of William I to the regency of Prussia marked a decisive move toward the abolition of the militia and its replacement by truly universal, far more effective conscription. The militia was unpopular with both the army's professional officers and its royal master because they saw it as lending undesirable muscle to the liberal and democratic forces in Prussia. The Landwehr's independence was therefore to be progressively diminished, until it became no more than an average reserve, retaining its traditional title but not a shred of its civic autonomy. The term of conscription, meanwhile, was to be reduced, and the obligation extended, so that all young Prussians—after 1870 all subjects of the Empire—were liable in their early twenties for intensive military training and, therefore, for recall on mobilization—with periodic refreshers in peacetime—until their late thirties.[11]

It was with this sort of short-service army that Prussia and the German states overthrew Napoleon III's France, which fielded an army raised by almost a parody of the system that prevailed in Prussia before 1857. Napoleon's army consisted of a standing army of conscripts committed to a term of seven years' service, reinforced by a citizen militia and the National Guard, both of almost no military value whatsoever.

In defeat, France drew from its painful experience the same conclusion that the Prussians had learned twenty years earlier: that military security and political safety alike demanded the creation of a short-service conscript army. It took the government of the Third Republic time to realize their mistake. The appalling civil disturbances provoked by the National Guard's attempts to seize power after the defeat of 1870, an episode we know as the Commune, encouraged the republic to persist in a long term of service—five years—until 1889. Thereafter service was reduced to three years and in 1905 to two. Admittedly it was increased again to three in 1913, but only because the high German birthrate had aroused fears that the size of the standing army was alarmingly small.[12]

Let us look for a moment at the conscription system in its heyday, taking as the exemplar the German army of 1913, while also examining the French and to a lesser extent the Austrian and Russian armies' adaptations of it. The German army comprised four elements: the Landsturm, the Active Army, the Reserve, and the Landwehr. At the age of seventeen a young German was registered for conscription and enlisted on the rolls of the Landsturm. At the age of twenty, if selected as fit, he passed to the Active Army for two years' service. He then moved on to the Reserve until he was twenty-eight, when he was transferred to the rolls of the Landwehr. Finally, until the age of forty-five, he was re-mustered into the Landsturm before returning finally and honorably to civilian life. Administratively he would have passed through the hands of a succession of units. At twenty he would have been enlisted in the unit with a permanent station nearest his home. If the unit was an infantry regiment, for example, it would have formed part of the local division and that, in turn, of the local corps; a corps was usually fielded by one of Germany's constituent states or kingdoms. Thus in 1914 Rommel's 124th Regiment formed part of the 27th Division; with the 26th it composed the XIII Corps, raised in Württemberg. The Reserve element in Würtemberg provided the 26th Reserve Division, with the older classes raising two Landwehr brigades and a number of Landsturm battalions.[13]

Würtemberg thus yielded to the Imperial German army a component at least three times larger than that the Royal Würtemberg Army had fielded as an independent force before 1870. But in this case there was more to the doctrine of mass than a mere accumulation of numbers. Two factors may be isolated in particular—the first material, the second moral. Materially, the transformation of the German—or French—army from a comparatively small long-service force into a large short-service force both demanded and was made possible by a transport revolution. Numbers make little difference unless they can be concentrated and deployed rapidly. Railways, which allowed the armies of 1914 to be moved at least six times faster than Napoleon's had been, made possible the creation of the mass armies of 1914.

Moreover, the availability of rail transportation for military purposes was far from accidental. The German railway system had been constructed under the direction of the German General Staff and according to a plan designed to serve the country's military as well as economic needs. Such a distinction is perhaps a tautology. Friedrich List (1789–1846), the principal theorist of railway economics in Germany and a dominant influence on the pattern of German railway building, explained that a wisely laid-out railway system "would enable the army of a unified Germany, in the event of invasion, to move troops from any point in the country to the frontiers in such a way as to multiply many fold its defensive potential.... [Moreover,] ten times stronger on the defense, Germany also would be ten times stronger on the attack." The layout List proposed in 1833, at the dawn of the railway era, coincided very closely with what was actually built; the offensive strategy he had foreseen but deprecated was ironically, because of the railroad construction, made both possible and desirable.[14]

But merely making mass armies transportable and provisionable would not have been enough to assure their effectiveness. A moral transformation of the spirit of the conscript contingent was a prerequisite if the mass armies of the late nineteenth century were to perform on the battlefield. And such a moral transformation did come to pass. Flight from the recruiting sergeant, a universal impulse in seventeenth-century Europe, gave way to a manly acceptance of the inevitable. How and why this change came about is unclear. In Germany it had a great deal to do with the carefully planned location of the units into which recruits were inducted; because a man left home not far behind him when he entered the barracks, home helped to ensure that he made a good soldier. But why nineteenth-century German households wanted their sons to become good soldiers is too large a subject to tackle here.

In France, the situation was different. Localization was deliberately eschewed (no French government wanted regiments of Parisians after the Commune). But a policy of using the army as the "school of the nation," deliberately and compensatorily pursued, achieved a wide measure of success. Dedicated to teaching the recruits "the Republican virtues," a euphemism for a sense of civic responsibility and national pride, the pre-1914 French army succeeded perhaps as well as the German in making young Frenchmen "think better of themselves for having been a soldier," to paraphrase Dr. Johnson.

By 1900 the legal, bureaucratic, logistic, and ethological processes necessary for the creation of mass armies had been stretched almost to their limits. On the one hand, in Russia and Austria–Hungary available resources of manpower were yet to be used to the fullest. On the other hand, in France, which conscripted 78 percent (and in 1913 would conscript 82 percent), the limits of these processes had probably been exceeded. Ironically, though, the liberation of manpower produced an embarrassment of riches that now faced planners with a new problem. There were almost too many men to be deployed usefully in the likely theaters of war. Ratios of men to space, at least along a future zone of Franco-German confrontation, approached five soldiers to each yard of front.[15]

For the French, committed to waging a war of concentric advance against Germany in concert with Russia, this overabundance of manpower did not matter greatly. For Germany, faced with the problem of defeating its enemies separately in order to overcome them collectively, the excess of manpower was a significant problem. There no longer being enough room to assemble and rail a mass army to the deployment area, space had to be found to maneuver the German force on the road network that lay behind the railhead. Schlieffen, Chief of the German Great General Staff from 1891 to 1906, labored for years seeking a solution to this problem. In the end he chose the expedient, but politically complicated, solution of extending his chosen front of attack beyond the limits of the Franco-German frontier and into neutral Belgium. As he explained in a 1901 strategic assessment:

> [The problem] grows simpler the stronger the enemy is, the further his lines extend, the more time it takes to support one attacked wing by the other one. How is the enemy's wing to be attacked? Not with one or two corps, but with one or two armies, and the march of those armies should be diverted, not against the flank, but against the enemy's line of retreat.... This leads immediately to ... disorder and confusion which gives an opportunity for a battle with inverted front, a battle of annihilation, a battle with an obstacle in the rear of the enemy.[16]

Students of blitzkrieg may detect in this analysis a foreshadowing of the practice the Germans would attempt in 1940–41. It may be judged ironic that the abstract consideration of operating mass armies had already led a protagonist to foresee limitations on their usefulness and thus to seek a means to apply force to a section rather than to the whole of the enemy's front.

But while the chief of the model mass army of Europe had already become a doubter of the efficacy of mere numbers, others who still lacked this wealth of resources strove to enlarge their own. That was most notably the case in France before 1914, when a static birthrate (about 70 percent of Germany's) showed signs of decline. For Colonel Charles Mangin the solution to the problem was the Empire. In his 1911 book *La Force noire*, Mangin, an officer of wide colonial experience, suggested increasing the number of African regiments in the French army and training them for European service.[17] A considerable proportion of the army was already North and West African; Turcos, as they were called, had fought in the war of 1870, and Moroccan, Algerian, and Tunisian tirailleurs would fight magnificently for France in 1914. Mangin's plea was for France to draw on its reservoir of black soldiers, particularly among the Senegalese, as a means of making good the widening gulf between French and Germans in Europe.

The French were not the only nation to make use of exotic sources of manpower. Russia, throughout its nineteenth-century period of eastward imperial expansion, had similarly enlarged its complement of cavalry regiments by recruiting from Cossack, Tartar, and Mongol border peoples. Austria had created its Muslim regiments from the manpower of Bosnia-Herzogovina after its annexation of those provinces in 1878, while the British policed much of their enormous overseas domain with its large Indian army.

Nevertheless, even in Britain with its powerful libertarian and anti-conscription tradition, demands for the introduction of compulsory service were heard well before the First World War. Lord Roberts, hero of the Second Afghan and Boer Wars and, later, leader of the movement favoring national service, found a powerful and persuasive supporter in Rudyard Kipling, whose strange story *The Army of a Dream* foreshadows a Britain in which the twin threats of invasion and social antagonism have been dispelled by the institution of universal military service.[18]

An Anglo-Saxon parallel to the continental preoccupation with conscription before 1914 is perhaps best exemplified in the populist rifle volunteer or sharpshooter movement that enthused middle and artisan classes all over northern Europe in the years between the mid-nineteenth century and the outbreak of the First World War. The best known of these movements was the British, which first appeared in 1859. By 1860 nearly 200,000 civilians had enrolled in rifle volunteer battalions. Tennyson's poem, "Form! Riflemen, Form!," both encouraged and celebrated the upsurge, and an unwilling government found itself obliged to regularize their status. Since the regular army wanted nothing to do with bands of shopkeepers in makeshift uniforms, for twenty years the Volunteers constituted a sort of armed Boy Scout Association, tolerated as well-meaning patriots but not taken seriously as defenders of the realm. The Volunteers' increasing efficiency eventually persuaded the government and army to integrate them into the state's military structure, and by 1908 the Territorial Force, as they were named, had come to constitute a true reserve. In 1914 the size of the Territorial Force was almost double that of the regular army; in 1915 it provided Britain with much of its Expeditionary Force in France.[19]

During this period similar citizen forces came into being in Belgium, Holland, and the Scandinavian countries, and we may find equivalents in the United States before the Civil War. The congruence of civic volunteering with democratic systems is certainly not accidental. Only in a self-confident democracy can the government feel sufficient trust in the benign intentions of voters necessary to allow them to acquire arms, drill, and train for war. Even so, the British volunteers incorporated an economic and class test as part of their admission policy. The French refused to admit the "dangerous classes"—unskilled workers, the unemployed, landless laborers. A blind spot in the legislation that regularized the British volunteers, however, permitted the raising of both the Ulster and Irish Volunteers in 1910-14, even though the aims of both bodies were sectarian and wholly inimical to civil order.[20]

By 1914, therefore, all the states that were to be combatants in the First World War had begun, in different ways, to make the mobilization of all available manpower the basis of their military systems. Some states, most notably France and Germany, had gone about as far in this respect as was possible. Indeed, neither France nor Germany was able to increase the real size of its army during the course of the war. By the end of 1914 Germany had about 80 infantry divisions in the field; by 1918 it had 241 divisions in existence on paper. But the total of effective formations was only some 160, a figure the Germans had achieved by reducing their fighting strength from 12 to 9 battalions. The French fared even less well. Their force of divisions increased from about 60 to only a little more than 100; again the increase was accomplished by splitting old divisions as a means of raising new ones.

Manpower flowed through these divisions, German and French alike, in enormous quantities. Indeed, by the beginning of the third year of the war some divisions had completely used up their original complement of infantry and were beginning to lose their replacements. By 1918 it was not uncommon for first-rate divisions to have suffered a 300 or even 400 percent loss of infantry. And by then no more replacements were left. Every sort of shift was employed to keep the divisions up to strength: Age limits were reduced and extended, casualties were returned to duty without proper convalescence, and—a breach of the ultimate taboo—the cavalry were dismounted and sent into the trenches. But all of this was to no avail. By mid-1918 Allied intelligence had categorized German divisions into four classes, the fourth of which was judged of virtually no military value whatsoever. Many Allied divisions found themselves in similarly dire straits. Only the Americans still had fit, enthusiastic young men in plenty. The mere knowledge that the Yanks were coming contributed significantly to the psychological collapse of the German army in September 1918.[21]

By the end of the war, evidence had accumulated that manpower was not enough, and the foundations of the manpower doctrine began to crumble. Although manpower was necessary in order to wage war, it was not sufficient to ensure victory. Economists might have warned military strategists of one permanent truth: When all restraints on the supply of a commodity are removed and investment of that commodity is entrusted to a monopoly, the monopoly will waste what it has been given. The enormous increase in the size of European populations during the nineteenth century had persuaded the generals that manpower supply was infinite. Consequently, they borrowed like spendthrifts, found that there was a bottom to the purse after all, and paid the price demanded of all spendthrifts.

A second permanent truth to emerge from the World War I experience is one that the political beliefs of post-revolutionary Europe had obscured: that not all men can be made into soldiers. Carnot's famous appeal had been based on the revolutionary idea that once men break their chains, their hands naturally reach to grasp a musket or sword. But many, perhaps a majority of men, like weapons no better than fetters. Warrior societies, so often technically backward and socially stagnant in comparison with trading and industrial societies, accept this concept as a principle of existence. The Germans in the Second World War, choosing to disregard this reality, persisted in efforts to make every man a soldier, and with remarkable success. The ferocity of their assault on the Soviet Union forced the Russians to act similarly. The British and Americans, by contrast, organized their armed forces using a much more discriminative process of selection that categorized men at induction as suitable or unsuitable for combat duty. The American decision to limit the size of its army to ninety divisions took into account the discriminative principle and certainly helped decrease American casualties.

Finally, the introduction of mechanization pulled away many of the props upon which the manpower doctrine had been erected. Not even the most headstrong military planner would argue that the elasticity of supply of tanks and aircraft was infinite. When machines, rather than men, became the cutting edge of armies, it was their availability that determined the useful size of a combat force. Tank crews, when killed, of course needed to be replaced. But since the loss of a crew usually also entailed that of the tank, machine and manpower replacement requirements were directly linked. Here we have a partial explanation for the often noted but inadequately explained decline of the operational size of armies during the twentieth century. Today we witness the result: states, richer and more populous than ever before, fielding operational forces half or even a quarter of the size of forces before 1914. The territory of West and East Germany, for example, which sustained a peacetime army of fifty-three infantry divisions in 1914, now supports eighteen modern equivalents.

Is the role of the mass army, then, over? Certainly, in the democracies this seems to be the case. Elsewhere the idea of a mass army persists. In the Soviet world the mass army continues to be seen as the most efficient means to discipline and train the young in their duties to the state. The idea exists as well in the developing world to the extent that states like Iraq, Iran, and Vietnam have yet to learn the misery and self-defeat that the effort to make every man a soldier necessarily brings in its wake.

NOTES

1. Christopher Duffy, *The Army of Frederick the Great* (New York: Hippocrene Books, 1974).
2. Christopher Duffy, *Frederick the Great* (Boston: Routledge & Kegan Paul, 1985).
3. Edward Mead Earle, ed., *Makers of Modern Strategy* (Princeton, N.J.: Princeton University Press, 1943), pp. 61-62.
4. Ibid., p. 98.
5. Samuel F. Scott, *The Response of the Royal Army in the French Revolution* (Oxford, England: Clarendon Press, 1978), p. 182; Jean-Paul Berthaud, *La Révolution armée* (Paris: Laffont, 1979), p. 271.
6. See, for example, D.G. Chandler, *The Campaigns of Napoleon* (New York: Macmillan, 1967), pp. 516-17.
7. Douglas Porch, *Army and Revolution: France 1815-1848* (Boston: Routledge & Kegan Paul, 1974), pp. 1-8.
8. Gordon Craig, *The Battle of Königgrätz* (London: Weidenfeld and Nicolson, 1964), chapter 1.
9. Michael Howard, *The Franco-Prussian War* (London: Rupert Hart-Davis, 1961), chapters 2 and 3.
10. See David Pipes, *Slave Soldiers and Islam* (New Haven, Conn.: Yale University Press, 1981).

11. Karl Demeter, *The German Officer Corps* (London: Weidenfeld and Nicolson, 1965), chapter 20.
12. David Ralston, *The Army of the Republic* (Cambridge, Mass.: M.I.T. Press, 1967), pp. 343-45.
13. *Histories of 251 Divisions of the German Army 1914-18* (Washington, D.C.: U.S. Government Printing Office, 1920), see entry on 26th Division.
14. Earle, *Makers of Modern Strategy*, p. 149.
15. Sewell Tyng, *The Campaign of the Marne* (London: Longmans, 1935), chapter 3.
16. See G. Ritter, *The Schlieffen Plan* (London: Oswald Wolff, 1958).
17. S.C. Davis, *Reservoirs of Men* (Westport, Conn.: Negro University Press, 1970).
18. Angus Wilson, *The Strange Ride of Rudyard Kipling* (London: Secker and Warburg, 1977), pp. 271-74.
19. See Hugh Cunningham, *The Volunteer Force* (London: Croom Helm, 1975).
20. See F.X. Martin, *The Irish Volunteers* (Dublin: J. Duffy, 1937); A.T.Q. Stewart, *The Ulster Crisis* (Winchester, Mass.: Faber & Faber, 1967).
21. See, for example, *The German Army Handbook April 1918* (London: Arms and Armour Press, 1977), chapter 3; J. Edmonds, *Military Operations France and Flanders 1918*, vol. 1 (New York: Macmillan, 1935), pp. 52-56; and P.M. de la Gorce, *The French Army* (London: Weidenfeld and Nicolson, 1963), pp. 93-119.

III THE CONTEMPORARY SITUATION

4 MILITARY MANPOWER IN CURRENT U.S. STRATEGIC PLANNING

Robert B. Pirie, Jr.

This paper deals with the question of how central a role manpower plays in U.S. strategic planning. In preparing this paper I did not have the benefit of other papers of a more theoretical bent to guide my inquiry. Lacking a framework, I have either to develop one or, in the extreme case, fall back on facts. So we begin with a central question: What evidence bears on whether manpower plays a central role in strategic planning?

Having fallen so low as to pursue actual facts, I will descend even further and consult my own experience. I was Principal Deputy Assistant Secretary of Defense (Manpower, Reserve Affairs and Logistics) from June 1977 to October 1978, the acting assistant secretary from October 1978 to June 1979, and assistant secretary from June 1979 to January 1981. Throughout that period I was actively involved in defense manpower policy and planning across an immense breadth of issues, including accession, retention, compensation, training, housing, medical care, retirement, force structure, mobilization planning, and many others. I was a member of the Defense Resources Board, the Defense Systems Acquisition Review Council, the Mobilization and Deployment Steering Committee, and the like. Thus I was in a position to observe and be involved in the development of such authoritative documents as the *Defense Guidance* and the secretary of defense's *Annual Report to the Congress.* Furthermore, it was my job to see that manpower received appropriate consideration in the development of policy and in the resolution of program and budget issues.

During these three and a half years, the issue of changing national strategy because of manpower constraints did not arise. There was significant concern as to whether and how we could man both active and reserve forces with volunteers. There was concern about whether the standby draft system could respond quickly enough in case of general mobilization. But the issue of changing strategy because of manpower problems never came up. Rather the reverse was the case. We were determined to man whatever forces the national resource allocation process ultimately mandated. We would produce volunteers if we could, or conscripts if we had to, but we would man the force.

A striking example of how manpower was not a prior consideration in strategy formulation was President Carter's 1980 State of the Union address, in which he declared the Persian Gulf an area of vital interest for the United States. The statement had obvious and immediate strategic implications, with the development of the Rapid Deployment Force one outcome. There was, however, no prior discussion, at least at my level, of the manpower implications of such a dramatic shift.

But perhaps the Carter administration was atypical or retarded in this regard. Perhaps others are more systematic. To discover whether or not this was the case I searched for evidence that the Reagan administration tests for manpower constraints in strategic planning. The initial findings were encouraging. The president, in a speech at West Point on May 27, 1981, said: "I have asked Secretary Weinberger to form a defense manpower task force to review the entire military manpower question...." The review would surely include whether different manpower policies would permit adoption of different strategies. But it did not. The task force limited itself to addressing the issue of whether the planned force could be sustained and mobilization requirements met.[1] Search of current posture statements and other such authoritative documents has also failed to disclose a link between strategic planning and manpower planning. I will later describe both in some detail, but for the moment it appears that in the past two administrations manpower considerations have not played a large role in strategic planning.

This, of course, is somewhat surprising. Population is thought by many to be an important determinant of military power. In contemporary books about strategy, manpower often gets top billing. For example, the recent book, *Strategic Requirements for the Army to the Year 2000*, states boldly that "manpower policy is a necessary element of an overall national strategy."[2] Similarly, in *Evolving Strategic Realities*, Franklin Margiotta devotes a subtitle and several pages to "Military Manpower as a Limiting Factor."[3] The problem is that these statements seem to reflect the author's or editor's views of what *ought* to be the case, not what is the case. Neither book makes a convincing case that manpower receives prior consideration as a significant determinant of national strategy.

We are left then with describing the actual process rather than the theoretical one. This is done admirably in the Department of Defense *Manpower Requirements Report FY 1986*, which says:

> Our wartime force structure is designed to satisfy our national strategy for mobilizing and prosecuting a war. The size and structure of that force and, in turn, the manpower requirements to man and sustain the force are determined annually in the wartime manpower planning system (WARMAPS) in the following sequential process.
>
> - The programmed combat force is prescribed in conceptual form in the Defense Guidance (DG). The DG also establishes a warfighting scenario. Each Service makes its plans and sets a variety of planning parameters, such as warning time and theaters of operations, based upon the DG.
> - Organizational, manning and stationing guidance is developed by each Service, promulgated in program documents, and supported, from year to year, in the Services' programs and budgets.
> - The mix and mission of force units between the active and reserve forces is determined and programmed by each Service according to its required contribution to the DG and/or JSCP [Joint Strategic Capabilities Plan].
> - All of these requirements compete with other priority programs for the limited resources available, and thus are constrained by authorizations and budget limitations.
> - Concurrent with the development of the combat force, the support establishment, which enables the committed forces to prosecute a successful war/contingency/engagement, is determined. Also, the cumulative casualty replacement requirements to sustain both the combat and support forces are determined.[4]

In other words, manpower is not an input to the strategic calculus; rather it is an output, or a residual.

What about feedback loops? That is, once a strategy has been selected and the forces, or their survivors, pulled through the knothole of the Planning, Programming and Budgeting System and congressional review, there must be a check to determine whether they can be manned. If the answer is no, then the strategy must be changed. In fact, this does not happen, nor has it happened. What does happen is that first a strategy is selected, then a set of forces is developed, less by reference to the strategy than to what we have inherited from the past and what we estimate we can support for the future. Then the system struggles to man these forces as a single, point requirement. If manning cannot be accomplished, we do not change the strategy, we have shortfalls. If the shortfalls are perceived to be big or serious enough, then the issue of a change in manpower policy arises.

What kind of shortfalls are a problem, and what kind of changes do people advocate? General Bernard W. Rogers, Supreme Allied Commander Europe, is so concerned about the shortage of U.S. Army reserve forces that he advocates a draft for the reserves. He is quoted as saying, "We're going to be short a minimum of 150,000 infantry, armor, artillery, [and] combat medics 90 days after a war starts in Europe."[5]

The *Manpower Requirements Report* also deals with this issue:

> When the supply of trained military manpower is compared with demand, we still find major shortfalls, particularly in the Army and Air Force. Because the timing of its peak demand and available supply is unique, each Service experiences a peak trained manpower shortfall at a different time. The Air Force's peak shortfall occurs at M+40 days, but the aggregate shortfall is eliminated rapidly thereafter; the Army's peak shortfall occurs at M+90 days, but is reduced rapidly as large numbers of volunteers and inductees complete training and join units. While the Marine Corps total trained manpower shortfall is eliminated by FY 1990, the combat enlisted shortfall increases steadily as the conflict progresses. In FY 1990, the Army faces a particularly large shortage of 111,000 in combat enlisted skills at M+90.[6]

However, the solution advocated by the report is not a return to peacetime conscription but an increase in the basic military service obligation from six to eight years. This increase would swell the size of the pool of trained individuals no longer in the army but liable to recall. Increasing the pool would presumably eliminate the shortfall.

It is at this point in virtually all such discussions of military manpower policy that the experts push their preferred solutions with gusto and cast suspicion on any alternatives. That is perhaps one reason why the strategy baby most often goes out with the manpower bathwater. In any case, the point is not which manpower fix is right. The point is that manpower considerations do not figure prominently, if at all, in the formulation of national strategy. Strategists view manpower experts in the same way they view logisticians—with extreme distaste. Once the strategy is decided upon, it is up to manpower to support it.

The reasons why manpower is not a principal determinant of, or constraint on, strategy are quite complex. To discuss them I shall have to invade, with some trepidation, the theoretical domains of other contributors to this book. My investigations have suggested four main reasons why manpower considerations do not play a strong role in the development of national strategy.

First, the overwhelming proportion of discussions of comparative military power, which determine strategy formulation and selection, are conducted in terms of force structures—numbers of army divisions, tactical aircraft wings and squadrons, and navy ships and battle groups. These force structures are not entirely independent of manpower considerations. One occasionally hears about concerns for manning a 600-ship fleet, for example, or, more dramatically, about hollow armies. But, by and large, our system has quite a bit of flexibility. The army has gone from 13 to 16 active force divisions and contemplates going on to perhaps 18, all within an active manpower limit of about 780,000. A major criticism of U.S. military style is that the support base seems to shrink less than the fighting forces when we phase down from military contingencies. While many observers are dissatisfied with our understanding of the efficiency and effectiveness aspects of tooth-to-tail trades, it appears that there exists the capacity to turn tail into tooth within manning constraints. Within broad bounds, then, discussions of what the United States can and cannot do with military forces tend to center not on how many soldiers we have but on how many divisions.

The second and related reason why manpower does not play explicitly in strategy formulation is that the resource allocation process works in a way to make dollars, not people, the binding constraint. This is clearly illustrated by the earlier quote from the *Defense Manpower Requirements Report* describing the planning process. Once again, manpower is a residual, not an input. Put another way, it makes little sense to bring into the armed forces more people than we can equip. All the military services, in spite of the recent good climate for defense budgets, have problems filling out their force structures with new equipment, not to mention buying spare parts and ammunition for sustainability. Indeed, in the last year or so the services have shown some inclination to limit personnel strength, despite a very good recruiting climate, in order to free money for higher priorities. This tendency will be reinforced by the adoption of accrual accounting for retirement, which increases the cost of an individual added to the force by 50 percent of base pay. This mode of prefunding retirement, which became fully effective in fiscal 1985, increases the marginal cost of additional personnel strength (and the savings from decreased personnel strength) and will, when fully internalized in the resource allocation process, make major differences in how the services manage military manpower.

At this point one can expect to hear the routine pitch for conscription. Why not, it is argued, save money by cutting military pay? If we do, we won't get any volunteers, so we'll have to draft the people we need. But we'll balance the books in such a way that we'll have larger forces, fully equipped with modern equipment, and fully supported with spares, ammunition, special tools, test equipment, and all the other good things. It is not the purpose of this paper to deal with that argument. The reason for bringing it up here is to address the point that manpower might be more of an explicit consideration in strategic planning if we changed the system that determines the price of manpower in a way that made people, not dollars, the constraint. Most serious studies of this issue have concluded, however, that there is not much money to be saved, because feasible reductions in pay for first-termers and savings in recruiting and advertising costs would be offset by increased training and other costs associated with high turnover and a smaller than desirable career/first-term ratio.

The third major reason for the poor convergence of strategy formulation and manpower planning is the fuzzy nature of strategy development. In principle one should be able to proceed from a statement of objectives, through an analysis of the impediments to achieving the objectives and an assessment of the costs and effectiveness of alternative means of achieving them, to a statement of the preferred force posture. In reality, however, it does not happen this way for a variety of reasons. First, the essence of strategy is secrecy. Much of what we intend to do and how we intend to do it is disguised in cryptic generalizations. And, secrecy or not, no president will be willing to sacrifice flexibility for precise statements concerning ends and means. Second, the process necessarily contains ambiguities and uncertainties that are irreducible and that, in many cases, dominate war outcomes. Examples are whether nuclear weapons will be used, and where and how.

A major uncertainty is the length of the war. Planning guidance for this is generally very ambiguous, and it is evident that, whatever the guidance, the various Defense Department organizations hedge their bets considerably. For example, if we are convinced that a war will be decided in its early days or weeks, it is hard to understand why we have such large reserve forces and maintain such a huge infrastructure of bases, depots, and the like. If, on the other hand, we expect any conflict between the great powers to be protracted, it is hard to understand the relative lack of emphasis on sustainability, the industrial base, and such long war necessities as sealift. In fact, if we faced the implications of protracted war, manpower might get much more explicit consideration in devising strategies. In a protracted war situation, we undoubtedly would have to call up a sufficiently large number of young people as to place a heavy burden on that portion of the population. Furthermore, with a real war going on, other resource constraints on force size would presumably be of less relative importance.

The fourth major reason why manpower does not affect strategic planning is the ambiguity and imprecision that attend the use of the word *strategy* in every quarter. Dictionary definitions tend to confine strategy to the science or art of military command as applied to the overall planning and conduct of large-scale combat operations. Clearly, this is too limited to explain modern usage, which takes into account business and sports strategies, weight-loss strategies, and the like in ways most people find intelligible. Nevertheless, some degree of limitation or subdivision of the use of the word *strategy* may be helpful. For example, one can think of three different, but not necessarily contradictory, meanings for "national strategy":

- A plan to defeat Warsaw Pact forces if they attack in Europe
- A plan to manage national resources to produce military forces capable of a wide variety of missions at acceptable long-run costs
- A plan to achieve national objectives in the long term by use of all available means—economic, political and military—which includes maintaining powerful military forces to deter war and exert political leverage

A common response to this point is, yes, we want all of these features in our strategy. Unfortunately, one can rarely get them all. Sometimes they conflict, and generally it is important to be specific about the strategy being addressed in designing programs and forces. The manpower policies appropriate to each of the above uses of strategy could be quite different. Thus, discussion of what manpower policies are appropriate to our national strategy can be unnecessarily bedeviled by confusion about what the objectives really are. General Rogers, intent on winning the war in Europe, advocates a manpower policy that he believes will get him the people he needs when he needs them. The secretary of defense and the president, however, are concerned with a much broader strategic formulation, in which conscription may have overriding negative effects or may simply be irrelevant.

Since use of the word *strategy* or the nature of the particular strategies under discussion seems to influence how manpower relates to strategy, it may be useful to examine a variety of strategies currently under discussion. For example, one strategy that appears often in the news and is strongly advocated by R. W. Komer is called "coalition warfare."[7] It advocates making as much use as possible of the fact that our allies are prosperous and populous and thus have strong incentives to defend themselves against Soviet coercion. Defense of Europe against Soviet attack will not be done by the United States alone but in concert with very large forces of allies. Therefore, our own forces and those of our allies must be designed to fight together. They must be able to communicate with each other at many levels, to execute plans and maneuvers in a similar and coordinated way, and to use each other's equipment, ammunition, and spare parts.

Implicit in this strategy, and often explicit in popular discussions of the matter, is the assumption that our allies need to do more in regard to their own defense—to share more of the burden. And the burden is characterized in terms of input resources and forces. This means our allies must provide additional manpower and thus relieve, to some extent, pressure on U.S. manpower. Indeed, some of the burden-sharing measures, most notably host nation support, are explicitly aimed at using allied manpower to perform a variety of tasks requiring non-military or minimally trained reserve personnel. Thus it can be argued that the strategy of coalition warfare has as a prime, if implicit, determinant an estimate of the U.S. manpower available for deployment in Europe during the early stages of hostilities and plans to augment that number with allied manpower.

Another strategic thrust that may be said to arise from manpower considerations is that of substituting capital for labor in military applications. One way of rephrasing the point is to say that we regularly use machines and advanced technology as force multipliers to offset manpower shortages or to reduce human risks and thereby compensate for our cultural aversion to casualties. This approach, advocated by many, including S. J. Deitchman,[8] has affected manpower policy in a variety of ways. The emphasis on recruits with high aptitude, for example, seems rooted in a perception that military service will have an increasingly technical and skill-demanding nature. The same can be said of the recent policy push to increase the career/first-term ratio. A more professional army is needed to deal with the complex equipment of war. The main problem is seen not as one of raising large numbers of minimally trained operators of simple systems, like rifles. Rather it is of creating a trained technical cadre and then keeping that cadre long enough to make adequate use of the time and training invested in it. The constraint, at least for the standing peacetime armed forces, and perhaps for the mobilization force as well, will be the amount of equipment that can be produced rather than the numbers of people who can be inducted and trained.

A strategic thrust much favored by reformers is the emphasis on maneuver warfare, and the consequent deemphasis of attrition as the prime deciding mechanism in war. This doctrinal reorientation has implications for manpower policy but its implementation is unlikely to result from or be determined by the manpower situation that presently confronts the United States. The implications, nonetheless, are quite interesting. Maneuver warfare depends on well-trained, intelligent, highly motivated, and adaptable troops. These are not the same bright people needed to maintain and operate the high-tech forces, however. Indeed, reformers tend to be very suspicious of high-tech solutions to military problems. The maneuver warfare forces are seen as being peopled by tough, enthusiastic warriors. These people are not attracted to the armed forces by economic or educational incentives. They are attracted to the profession of arms. By some further transformation of logic that I am not able to replicate, an army of conscripts is thought to be more likely than a volunteer army to attract such people. However, whatever the logic of the argument, this is a case in which the manpower requirements are an output of, rather than an input to, the process of strategy formulation.

A strategy for achieving maximum deterrent effect from limited resources may be influenced by perceptions of manpower constraints and may have further consequences for manpower policy. Current U.S. strategy appears to be an example, although as I have argued previously, dollar constraints seem more binding than manpower constraints. In any case, U.S. policy is to maintain large and formidable active forces, within current resource constraints, as well as very substantial reserve forces, which provide a mobilization base and offset, in part, the fact that the active forces are not as large as we would like. In the army the large active force structure is made possible within total available manpower by placing a large amount of the support that would be required in war for the active force in the selected reserves. James Lacy, a former member of the Atlantic Council's Working Group on Military Service, has argued that it is manpower constraints that have forced this upon us.[9] Indeed, Lacy points out that we have relegated active functions to the selected reserve, selected reserve functions to the individual ready reserve (IRR), and IRR functions to mobilization draftees, and at each step we have accepted reduced readiness as an accommodation for limited resources. Whether this is wise or whether manpower is the dominant constraint is beside the point. The fact is that there are, in this case, strong interactions between strategy and manpower policy.

The point is that there are circumstances in which manpower is or could be a prime determinant of strategy, and not a residual of a predetermined course of action. Under contemporary conditions this apparently does not happen. It has not happened since World War II, if it happened then. The fundamental reason for this appears to be that the United States has an abundance of young people available at market prices for military service, especially in comparison with the other resources American taxpayers are willing to provide for the common defense. Constraints that are considered in the formulation and articulation of national strategy tend not to arise from manpower issues. This leads one to the ineluctable conclusion that manpower does not play a central role in current U.S. strategic planning.

NOTES

1. *Military Manpower Task Force: A Report to the President on the Status and Prospects of the All-Volunteer Force* (Washington, D.C.: U.S. Government Printing Office, November 1982).
2. Robert H. Kupperman and William J. Taylor, eds., *Strategic Requirements for the Army to the Year 2000* (Lexington, Mass.: Lexington Books, 1984), p. 143.
3. Franklin D. Margiotta, ed., *Evolving Strategic Realities: Implications for U.S. Policymakers* (Washington, D.C.: National Defense University Press, 1980), esp. pp. 190–94.
4. U.S. Department of Defense, *Manpower Requirements Report FY 1986*, vol. 3, *Force Readiness Report* (Washington, D.C.: Department of Defense, February 1985), p. II-12.
5. *Washington Post*, 2 March 1985, p. 7.
6. U.S. Department of Defense, *Manpower Requirements Report FY 1986*, p. II-14.
7. Robert W. Komer, *Maritime Strategy or Coalition Defense?* (Cambridge, Mass.: Abt Books, 1984).
8. Seymour J. Deitchman, *Military Power and the Advance of Technology* (Boulder, Colo.: Westview Press, 1983).
9. Andrew J. Goodpaster, Lloyd H. Elliot, and J. Allan Hovey, Jr., *Toward a Consensus on Military Service: Report of the Atlantic Council's Working Group on Military Service* (Elmsford, N.Y.: Pergamon Press, 1982), p. 200.

5 STRATEGIC INFLUENCES IN MILITARY MANPOWER PLANNING

Sam C. Sarkesian

An analysis of the capabilities and fighting qualities of soldiers involved in the Normandy invasion is part of an absorbing account of Operation Overlord by Max Hastings. In one case, a British officer is quoted as saying, "The trouble with our British lads is that they are not killers by nature."[1] In another instance, a British brigadier stated, "We are always aware of the doctrine, 'Let metal do it rather than flesh.' The morale of our troops depended upon this. We always said—'Waste all the ammunition you like, but not lives.' "[2]

The American army also had its problems in developing fighting instincts in its soldiers, while leaning heavily upon technology and firepower to achieve military objectives. A particularly serious indictment of American manpower policies was Hastings' observation that " . . . the American army's 'teeth' elements were severely blunted because they lacked their proper share of the ablest and fittest officers and men."[3] He noted that in the American army,

> The air corps, the specialist branches, and the service staffs had been allowed to cream off too high a proportion of the best-educated, fittest recruits. Infantry line companies would be called upon to fight Hitler's Wehrmacht, "the most professional skillful army of modern times," with men who were, in all too many cases, the least impressive material America had summoned to the colours.[4]

Thirty years and two wars later, General Fred C. Weyand, analyzing the Vietnam War, stated, "There is no such thing as war fought on the cheap. War is death and destruction. The American way of war is particularly violent, deadly and dreadful. We believe in using 'things'—artillery, bombs, massive firepower—in order to conserve our soldiers' lives."[5]

In addressing the general issue of combat effectiveness in the contemporary period, Morris Janowitz points out that the military has become a constabulary force more attuned to functioning as a political-psychological instrument than as a combat force. This is a result not only of deterrence policy but also of the difficulty of maintaining a combat orientation without periodically sharpening the combat edge in real combat situations.[6] While it may be difficult to demonstrate the relationship between a fighting ethic and deterrence policy, it is less difficult to link the fighting ethic to the nature and character of society. "Social changes will make it more rather than less difficult to reconcile the needs of the services for disciplined and well-motivated manpower with an increasingly libertarian and anarchic society."[7] In brief, the concept of a constabulary force in combination with the nature of American society has produced conditions that erode the traditional raison d'être of the military.

The general thrust of Janowitz's arguments appears to be supported by the view that an occupational orientation has emerged in the volunteer military system. Moskos' work on occupational-professional issues indicates a growing occupational orientation within the military that closely parallels the civilian job ethic and mind set, rather than the fighting ethic of the traditional military.[8]

In sum, in the Normandy campaign, soldiers from Western democracies tended to rely heavily on technology and massive firepower to engage the enemy. While this was a distinct advantage in the overall campaign, at lower operational levels these doctrines tended to minimize unit initiative and individual enterprise. The technical orientation and high regard for specialists led to a concentration of the best and the brightest in the American military in non-front line units and duties. Studies of the American military in the years since World War II indicate that these technological, managerial, and specialist tendencies have increased. The notion that American society, like other Western systems, does not inculcate a serious fighting ethic in individuals forms the basis for most of the observations that have flowed from these studies.

Furthermore, deterrence policy, combined with the volunteer system and its tendency to highlight an occupational model of service, has eroded the traditional view of military service and may be harmful to combat efficiency and the fighting spirit; at the very least, it may have blunted the sharp combat edge necessary for success in battle. What has exacerbated the problem of manpower-strategy congruence is the persistent fear that in the long run the volunteer system will be unable to provide the manpower levels necessary to carry out a global strategy, much less to sustain major wars or even limited wars of some duration.

Finally, historically the American way of war has been based on the assumption that wars are initiated by a clear outbreak of hostilities between two powers, with the issues clear, the enemy identifiable, and the ultimate goal total victory. All of these factors have serious implications with respect to the fighting ethic and, moreover, in terms of political-military policy, strategy, and manpower planning in the current period. In the most extreme interpretation, these developments have not only eroded the combat effectiveness of the military but they have negatively affected civilian views of military capability and, in turn, impaired the national will and the political resolve of elected officials to use military force in all but the most threatening contingencies.

While these observations are generalizations that must be qualified, the fact is that strategy and manpower seem to be oriented toward the highest intensity type wars—modern versions of World War II—or toward limited conventional conflicts for which forces-in-being are presumed adequate. Thus, on the one hand, manpower planners focus on mobilization-type wars, which entail few restraints on manpower levels. On the other hand, these same manpower planners concentrate on maintaining extant force levels, presuming that such levels are adequate for implementing current strategy. Regardless of the contingency planning, the underlying strategic orientation rests on conventional scenarios, emanating from a Eurocentric focus on battles in Europe against a Soviet adversary.

What do these observations mean with respect to strategic and manpower issues in the contemporary period? In brief, each of these observations, from that of the Normandy campaign to deterrence, reflects major issues of the manpower-strategy relationship. Additionally, these observations point out the differences among the perspectives of manpower analysts, strategic planners, and conflict contingencies or among numbers, missions, and conflict realities. In brief, conventional perspectives pervade strategic outlooks and in turn determine manpower policy. The result is a concentration on the least likely scenarios, i.e., those interpreted in conventional terms. The projected scenarios have their parallels in manpower policy.

This chapter analyzes the relationships between strategy and strategic concepts on the one hand, and military manpower composition and planning on the other hand, to determine how these relationships affect the most likely contingencies facing the United States in the coming decades. This analysis will identify the mismatches between these various components and develop guidelines for a reformulation of strategy and manpower planning. The intent is not to design an operational directive or to correlate programs linking manpower procurement to conflict requirements. Rather, the intent is to advance the proposition that the nature of modern conflicts necessitates a closer integration of strategy and manpower perspectives. Such perspectives must seriously recognize the existence of a "new" battlefield and examine what this new battlefield means in terms of manpower policy and planning.

THE STRATEGIC PERSPECTIVE

Although the term *strategy* is used frequently in studying the military posture and policy of various states, in American military thought it is the Clausewitzian notion that dominates: "Strategy is the employment of the battle to gain the end of the War; it must therefore give an aim to the whole military action, which must be in accordance with the object of the War."[9]

A further interpretation is provided by General Bruce Palmer:

> The term "strategy," derived from the ancient Greek, originally pertained to the art of generalship or high command. In modern times, "grand strategy" has come into use to describe the overall defense plans of a nation or coalition of nations. Since the mid-twentieth century, "national strategy" has attained wide usage, meaning the coordinated employment of the total resources of a nation to achieve its national objectives.[10]

But as British historian Michael Howard points out, "[t]he term 'strategy' needs continual definition. For most people, Clausewitz's formulation ... is clear enough. Strategy concerns the deployment and use of armed forces to attain a given political objective."[11] The utility of the modern state and conventional views regarding the relationship of war and politics are essential parts of strategy. This is the kind of formulation that guides the American military, even though there are those who challenge the utility of war as an instrument of policy, particularly as an instrument of American policy. As Russell Weigley has remarked, "At no point on the spectrum of violence does the use of combat offer much promise for the United States today.... Because the record of nonnuclear limited war in obtaining acceptable decisions at tolerable cost is also scarcely heartening, the history of usable combat may at last be reaching its end."[12]

Nonetheless, military strategy in its most simple definition retains as its centerpiece nuclear weapons and combat envisioned between major powers. The Joint Chiefs of Staff have affirmed that "the fundamental ingredients of U.S. military strategy are nuclear deterrence with arms control, strong alliances, forward-deployed forces, central reserves, force mobility, freedom of the seas and space, command and control, and intelligence."[13]

Yet, there is a school of thought within the military that recognizes the challenge of unconventional conflicts. Not only was a Special Operations Command created in recent years, but increasing resources have been made available to special operations forces. Additionally, special operations career patterns have been created within the American army and a series of initiatives undertaken to increase overall special operations capabilities. In terms of manpower and financial resources, however, the amounts allocated to special operations forces are miniscule, even though they represent quantum jumps from allocations in previous years.[14]

What does all of this mean with respect to strategists and strategy? Primary attention remains on nuclear strategy and wars of a global nature. While some attention has been given to limited wars, they tend to be seen merely as conventional conflicts in special settings. Accordingly, military men and women, properly trained in the traditional sense, presumably can perform well in virtually any set of circumstances. Moreover, there also is a presumption that military professionalism and prevailing military organizations and doctrines are appropriate for all but the most "unique" kinds of combat. Complicating the relationship between manpower and strategy is the fact that among civilian strategists there are some sharp disagreements and much debate over force structures and American security interests.[15]

Not only do traditional strategic formulations shape the debate, but to a great extent they also determine organizational philosophy and give substance to military professionalism. Conventional views of conflicts, that is, battles between major powers or between similarly postured foes, underpin doctrine and military training. There is a tendency, therefore, to concentrate on manpower levels based on assumptions about the nature of conflicts that emanate from strategy based on Clausewitzian notions of military objectives and missions.

It is normal practice for strategists to design scenarios from which manpower planners formulate plans to acquire the personnel necessary to sustain various contingencies. This presumes that manpower planners have some understanding of the nature and character of various conflicts in addition to skills in manpower planning. Although this may be the case with respect to major conventional and nuclear war contingencies, it is less so with respect to other types of conflicts. Major conflicts are expected to make the greatest demands on troop strength. Also, these are the kinds of conflicts that are likely to precipitate and sustain national will and political resolve. Hence, mobilization undergirded by the impulse to protect the American homeland is likely to engender few constraints on manpower levels. There is a considerable mismatch, however, between strategy and manpower policy in terms of response, force structures, and requirements for small wars and unconventional conflicts. These conflicts are not likely to generate the national will and political resolve needed for prolonged conflict.

Complicating and confusing the matter is the fact that strategy and manpower must be conceived within the general outlines of policy and national objectives. Policy needs to provide guidelines for identifying American interests and goals as well as the conditions under which political-military forces will be used to achieve these particular goals. If policy is unclear or ambiguous, strategists will tend to identify and interpret policy and national objectives in the way that best fits their own strategic perceptions. Not only can this lead to inconsistency and contradictory strategic options, it also can create major gaps between strategy and manpower requirements. In this respect, while there is disagreement regarding the policy of deterrence and the use of force in general, the fact is that in strategic terms the identification of the adversary and his military capability are reasonably clear. But even here there is some concern about policy clarity and coherence. The larger policy problem has to do with conflicts of lesser magnitude. This is particularly the case with respect to unconventional conflicts and the use of special operations forces. These policy problems have their parallels in strategy and manpower issues.[16]

MANPOWER PLANNING

To say that there is considerable debate about manpower needs—recruitment, retention, and quality—in the volunteer system is to state the obvious. There are a number of published volumes and a variety of in-house studies on all of these issues.[17] More often than not, these studies focus on the economics of the marketplace, incentives for recruitment and retention, and the quality of personnel in terms of education and age. With increasing frequency, questions are being raised concerning social representativeness and the role of women in the military.

The general arguments for or against various manpower policies are well known and need not be repeated here. It may be useful, nonetheless, to briefly review one of the more challenging aspects of the debate—the volunteer system and its alternatives. Three positions have been advanced, all based on the presumption of a declining youth population. The first position supports the volunteer system and seeks to increase its effectiveness by expanding the role of women and enlarging recruiting incentives, among other things. The second position advocates conscription with all of its well-known characteristics. A third alternative is national service, which gives individuals an opportunity to select from a variety of options to fulfill their obligations, with the military being only one of many options.[18]

This brief overview does not reflect the complex political, social, economic, and budgetary issues associated with various manpower options. But the point is that many fear that future military manpower requirements are not likely to be met by the present volunteer system. These individuals argue that adjustments are necessary and alternatives need to be considered. While these are important issues, one of the most important issues tends to be lost in the debate: the link that properly should exist among manpower, strategy, and conflict contingencies.

Whether strategy and conflict contingencies should be a concern of manpower planners may be argued by some. The fact is, however, that manpower planning devoid of strategic considerations is reduced to a sterile proposition where numbers become the driving force, subordinating or ignoring factors that affect the use and motivation of personnel. According to two experts,

> Manpower policy is a necessary element of an overall national strategy. Factors that affect the number and quality of military forces, from privates to three-star generals, are inevitably derived from the doctrine for the use of those military forces. A successful policy will satisfy a military requirement for tough, intelligent combat troops in sufficient numbers to prevail in combat, the political need for a punch of sufficient credibility to deter aggression, the fiscal need to hold down standing forces that contribute nothing to the gross national product, and social concerns about the just use of force in the modern world.[19]

This is a tall order indeed. Not only are manpower planners saddled with the task of identifying troop levels and sources of recruits, they must understand national strategy as well. Conversely, strategists must surely include as part of their calculations manpower availability, quality, and socialization toward military service and conflicts in general.

Yet the general thrust of current manpower studies and planning is conditioned by factors that appear to be driven by marketplace economics and demographics, with minimum regard for the more subjective and political–psychological factors. A more comprehensive and strategically oriented military manpower policy must go beyond *general* quality, quantity, and retention factors. To link manpower and strategy, the manpower planner needs to develop a policy template based on a number of variables.

> ... First, he must project the types of missions his forces will be asked to perform.... Second, he must postulate the constraints that will be placed on such military actions.... Third, he must work realistically within the limitation of the recruitment, retention, and reserve capacity of the various military branches. Fourth, he must project the abilities of the various recruits to operate and maintain increasingly sophisticated equipment in unconventional battlefield environments. Finally, he must forecast... categories that the military will require.[20]

Unfortunately, there is little serious analysis of these variables in the existing literature, although there are some notable exceptions such as the works of Janowitz, Wesbrook, and a recent study on military cohesion by Henderson.[21]

Additionally, much current manpower policy and planning rests on traditional notions of strategy that view conflicts from conventional perspectives and focus on those that are least divisive of American society. In brief, the big-battle syndrome of the European continent and nuclear deterrence policy are the strategic linchpins of manpower policy. Even when some attention is given to other scenarios, conventional thought patterns dominate within the context of traditional strategic boundaries.

Some authorities, however, challenge prevailing assumptions regarding global strategy and force requirements.

> ... Overall manpower requirements for the "Total Force," are in turn shaped to a considerable degree by the U.S. commitment to NATO, as well as anticipated demands on U.S. general-purpose ground forces should a conflict in Europe occur.... But are such forces necessary? The United States is a global power with global interests, it must therefore have at its disposal a military establishment whose size, composition and "strategic reach" are adequate to secure those interests. What constitutes "adequate" force, however, is far from certain.[22]

The strategy–manpower mismatch created by these policy and strategy outlooks has multi-dimensional implications. Not only are there problems in identifying and nurturing the fighting ethic in the contemporary conflict environment, but there are problems associated with establishing and maintaining force levels and force structures in support of American global interests.

It is possible that a clear threat aimed at the heart of the American homeland can overcome most problems associated with force levels and the fighting ethic. For example, one authority concludes,

> Even if deterrence is acknowledged to affect the fighting ethic, there is little ground for imputing either positive or negative connotations to the relationship.... Furthermore, it is questionable that deterrence would significantly degrade one's ability to defend the homeland. It is only where "defense" involves ambiguous, or questionable, political objectives that there are potential problems.[23]

But in the contemporary period, the conflicts most likely to threaten seriously American interests are those that have "ambiguous, questionable political objectives."

SUMMARY: STRATEGY AND MANPOWER

The function and tasks of the strategist and those of the manpower planner converge in mobilization-type wars. At this level, strategists and manpower planners may be asking the same kinds of questions, such as, What levels of troop strength are needed to maintain the force levels necessary to engage successfully in these conflicts? In this context, the strategy-driven nature of manpower planning, the conventional mind-sets, and the traditional strategic orientation dominate the military environment. Issues of socialization, attitudes toward war, national will, staying power, and the fighting ethic are infrequently incorporated into strategic and manpower equations.

The necessary correlation between manpower and strategy is contingent upon developing strategic appreciation by manpower planners and manpower appreciation by strategists. To a great extent this is best done at the major war level, where other political-psychological matters are not so pressing. The more serious strategic problem, and hence manpower problem, lies in addressing conflict contingencies at the lower end of the spectrum. The serious problems facing manpower planners are not only designing manpower programs for various conflict contingencies, but providing clear strategic guidelines at the lower end of the conflict spectrum. Indeed, strategy may even misguide manpower planners and policymakers in that arena.

The fundamental problem is that conflicts of a lesser magnitude—unconventional conflicts—pose a series of manpower problems that differ considerably from those associated with the big-battle scenarios, conventional perceptions, and traditional strategic thought. The largest gaps and mismatches between strategy and manpower arise primarily from the nature and character of unconventional conflicts and American policy. These gaps and mismatches are clearer when comparing the strategic posture and the manpower base with the actualities of conflicts along the entire conflict spectrum. Not only does such a comparison highlight the strategy/manpower problem, but it shows conceptual and capability mismatches specifically linked to that problem (see Figure 5-1).

Strategic formulations rest, in part, on perceptions of threat and views regarding the most likely adversaries. The American posture is shown in Figure 5-1. The scope of each category is intended to show the amount of attention and resource allocation for each category.

Figure 5-1. The Conflict Spectrum.

Noncombat	Unconventional Conflicts	Conventional	Nuclear
Shows of Force	Special Operations	Limited Major	
Assistance	Surgical		
	Hit and Run		
	Counterterror		
	Spearhead		
	Rescue		
	Counterinsurgency		

U.S. Concentration of Effort and Allocation of Resources

⟵——————————————————————————⟶

Low	High

Thus, at the two extremes, there is a great deal of political–military attention and large allocations of resources. These are likely to continue.

The lowest end of the conflict spectrum is characterized by a variety of noncombat situations ranging from low levels of military assistance to shows of force. The high end of the spectrum is characterized by big-battle scenarios, both conventional and nuclear. In the middle range are unconventional conflicts. These categorizations are intended to show policy and strategy perspectives and are not intended to delineate the degree of conflict intensity within each category.

The conflict spectrum shown in Figure 5-1 is not meant to convey the idea that clear boundaries separate various types of conflict. That is, at times it is difficult to know, for example, when unconventional wars become conventional and when limited, conventional wars become major conventional wars. However, the categories do place a premium on the *critical thrust* of each type of conflict. In this respect, unconventional conflicts have characteristics that clearly distinguish them from conventional ones. This has a number of implications for American policy, strategy, and manpower requirements that differ considerably from conventional conflict requirements. In terms of manpower, these differences go beyond quantity and general quality of personnel.

The dilemmas and problems are magnified if a more realistic conflict spectrum is examined.[24] Figure 5-2 shows the conflict spectrum

Figure 5-2. The Conflict Spectrum—Revised.

Noncombat	Unconventional Conflicts		Conventional	Nuclear
	Special Operations	Low-Intensity Conflicts		
Shows of Force Assistance	Surgical Hit and Run Counterterror Spearhead Rescue	Revolution Counterrevolution	Limited Major	

Level of Intensity

◄───►
Low   High

U.S. Concentration of Effort and Allocation of Resources

◄───►
Low   High

Current U.S. Capability

◄───►
| Adequate to Good | Poor to Adequate | Adequate to Good |

as it should be conceptualized for the contemporary period and the years ahead. As Michael Howard observes, "Most strategic scenarios today are based on the least probable of political circumstances—a totally unprovoked military assault by the Soviet Union, with no shadow of justification, on Western Europe."[25] The point is that the vast middle range of conflicts is the most likely in the foreseeable future, and it is in this range that the United States appears to be the least prepared in terms of policy, strategy, and manpower planning.

The strategists' attention to unconventional conflicts has been incidental to their preoccupation with the high-intensity end of the conflict spectrum. Attention given to unconventional conflicts has either been totally in terms of the Vietnam experience or seen through conventional lenses and traditional strategy. Thus, the Clausewitzian notion of war and battle has shaped the American response to unconventional conflicts. To be sure, conventional scenarios are important ingredients of American strategy, but they are not the only ones. Indeed, other scenarios may be more important in light of the contemporary international security situation.

Even in the big-battle scenarios, however, many disagree over policy and strategy. This has important implications for manpower policy and planning.

> The military establishment, and particularly general-purpose forces, must be reappraised with a view to the type of manpower demands future conflicts are likely to place upon them. This requires a careful evaluation of the missions the armed forces must be able to perform in support of U.S. foreign policy.[26]

Conceptual confusion over the meaning of unconventional conflicts has added to inadequate strategic analysis and, just as important, reflects a great deal of policy ambiguity. These problems are made clearer in comparing Figures 5-1 and 5-2. The gaps between strategic scenarios and strategic realities are obvious.

Manpower planning reflects the same gaps and problems, paralleling strategic perceptions. This is exacerbated by prevailing concepts of military training, education, and general professional posture. These concepts are based on the view that military men and women trained in conventional warfare and skilled in the basics of individual soldiering are capable of adapting to all types of conflicts. Thus, manpower quantity and quality remain wedded to conventional and, implicitly, mobilization-type conflict planning assumptions.

THE AMERICAN POLITICAL SYSTEM: IMPACT ON STRATEGY AND MANPOWER POLICY

Strategic considerations and manpower planning take place within the context of democratic values, expectations, and the dynamics of the American political system. There is little need to review these characteristics. Suffice it to say, the dynamism and pluralism of the system generally preclude the establishment of a single office with unlimited power to plan, train, and commit American military forces. The dynamics between the executive branch and Congress are well known. Moreover, American national will and political resolve evolve, to a large degree, from the perceptions of the American people regarding the commitment and capability of American forces.

It follows that American staying power rests, in general, on the public's acceptance of proper courses of military action and the right policy and strategy. This staying power must be nurtured and sustained over a period of time—it cannot be a one-time phenomenon. And as history has shown, in a democracy it is difficult to develop and maintain staying power in prolonged conflicts, particularly those that are ambiguous and politically questionable. Conflicts with clearly identifiable enemies and purposeful goals, like those evolving out of Pearl Harbor events, generate a staying power that is considerably different than that generated by conflicts like Vietnam or even Korea.

General Weyand has pointed this out clearly with respect to the American army and Vietnam.

> Vietnam was a reaffirmation of the peculiar relationship between the American Army and the American people. The American Army is a people's Army in the sense that it belongs to the American people who take a jealous and proprietary interest in its involvement. When the Army is committed the American people are committed, when the American people lose their commitment it is futile to try to keep the Army committed. In the final analysis, the American Army is not so much an arm of the Executive Branch as it is an arm of the American people. The Army, therefore, cannot be committed lightly.[27]

Affirming the importance of this army-people relationship, General Weyand charged the military professional with an important political-psychological mission.

> As military professionals we must speak out, we must counsel our political leaders and alert the American public that there is no such thing as a "splendid little war." ... The Army must make the price of involvement clear *before* we get involved, so that America can weigh the probable costs of involvement against the dangers of noninvolvement . . . for there are worse things than war.[28]

Although General Weyand directed these statements at the American army, their relevance to the military institution as a whole is clear.

In the aftermath of Vietnam, Congress has taken a more assertive role in military matters, and a variety of other political actors have become deeply involved in the debate over defense expenditures and military involvement. Indeed, in some instances, even the hint of American military involvement causes a nervous reaction within the body politic (e.g., aid to the Contras in Nicaragua). Equally important, as suggested by General Weyand's comments, the military system itself has become sensitive, some argue supersensitive, to domestic political reactions to military involvement.

While the volunteer system eased some of these problems, it has not provided much latitude in committing and maintaining military forces outside of standing treaty obligations such as NATO. Moreover, the constant and pressing issues of recruitment, retention, and defense budgets perpetuate a relatively high military profile. Fears of another Vietnam, reaction against the executive office resulting from the Watergate affair, the high military profile, and the sensitization of a number of political actors in and out of government have created a domestic–political environment in the United States that places a number of limitations on the use of military force, except in the most threatening circumstances.

While this situation does not preclude attention to scenarios across the entire conflict spectrum, it does create a chilling effect on the serious analysis of, and attention to, the use of military forces in all but either high-intensity conflicts or low-visibility conflicts at the other end of the spectrum. In other words, the American military institution, in particular the army, has become national in the sense that it is expected to adhere to the views and attitudes of the American people in regard to military commitments. Moreover, most Americans expect the behavior of military men and women to adhere closely to democratic norms.

These conditions are particularly relevant in scenarios that presume visible employment of conventional military forces, as in Grenada. The situation may well be different in scenarios that require low-visibility American involvement, such as the commitment of a special forces team, or where small detachments of special operations forces engage in quick strike, withdrawal, or hostage-rescue missions.

GUIDELINES FOR THE FUTURE

Conflict contingencies in the contemporary period and through the coming decade suggest that U.S. strategy and manpower policy must be reexamined and related more closely to the conflict spectrum. Equally important, this must be done in the context of constraints and limitations imposed by the American political system.

Strategic Guidelines: New Dimensions

Strategy conceived in traditional terms may need to be reformulated to correspond more closely to the nature and character of contemporary conflicts, especially those most likely to occur in the immediate future. It follows that strategists need to devote increasing attention to factors that have not normally been included as part of the strategic equation. According to Michael Howard, "The historical conditions which made traditional strategy possible have now very largely disappeared.... The whole governmental control of violence which made military strategy possible at all... is being rapidly eroded. The state monopoly has been broken."[29]

Howard also challenges Western interpretations of Clausewitzian notions concerning the nature of strategy. Strategy, Howard argues, includes "... four dimensions: the *operational*, the *logistical*, the *social*, and the *technological*."[30] He discusses these dimensions of strategy as evidenced from the American Civil War through the contemporary period, points out their relationships, and concludes that they have been repeatedly neglected or confused by Western strategists.

> Works about nuclear war and deterrence normally treat their topic as an activity taking place almost entirely in the technological dimension. From their writings not only socio–political but the operational elements have quite disappeared.... Drained of political, social and operational content, such works resemble rather the studies of the eighteenth-century theorists whom Clausewitz was writing to confute, and whose influence he considered, with good reason, to have been so disastrous for his own times.[31]

Following this critical analysis of traditional strategy, Howard argues that the socio–political dimension has become extremely important in current strategy. This is particularly true with respect to the nature and character of conflicts that the United States and other Western nations are most likely to face in the coming years—those in the middle range of the spectrum (see Figure 5-2). These conflicts have more to do with the political–psychological and social dimensions of strategy than with the actual "battle." While armed conflict is important in such cases, it is not likely to be the overriding strategic focus. That is, the center of gravity of such conflicts may not be the enemy's armed forces, but rather the political-social milieu of the indigenous system. Political cadre, political organizers, and psychological warfare specialists may be more important than combat personnel. Sun Tzu's notion of moral influence and winning without combat may well be a more realistic strategic focus than the Clausewitzian emphasis on destroying the enemy's army.[32]

The social dimension of strategy refers not only to the sociopolitical environment of the target area and the adversary but to the socio-political cohesion of the employing power. This is not to suggest that other elements of strategy are unimportant. But whereas the United States may have excelled in the technical, logistical, and operational dimensions of *traditional* strategy in Vietnam, these were not enough to overcome the strategic impact of the broader socio-political dimension.

If this is the case, the need for forces designed for mobilization-type wars and major conventional conflicts, while still important, may be less compelling in terms of manpower policies and strategic imperatives. Such forces may be more important in terms of deterrence than actual combat. It may well be that unconventional conflicts—small wars (in which special operations forces are important) and counterterrorist operations—will assume far more important roles in any strategic equation. It follows that special operations forces will be more important for the foreseeable future than massive armies or high levels of forces-in-being. This does not necessarily mean decreasing the importance of forces-in-being as presently conceived. What it does mean is a major strategic reorientation toward special units and unconventional conflicts with a concomitant shift in manpower planning. Thus, in designing these smaller forces, manpower planners should concentrate on identifying and recruiting individuals who are most success oriented and more likely to develop and maintain the fighting ethic.

Furthermore, it is conceivable that present force levels and strategy aimed at deterrence and conventional conflicts are becoming increasingly inappropriate. Air mobility, sophisticated weaponry, and alliance political-military policies and sensitivities, for example, may necessitate different American political-military postures, strategies, manpower policies, and force levels.[33]

The strategic equation should therefore seriously address a variety of questions that tend to be overlooked in traditional strategic analysis: What conflicts are most likely to occur in the contemporary period? How are these likely to affect American national interests? What strategic options are available to respond effectively to these conflicts in a period of manpower constraints? What forces are required to support these strategic options? What quality and composition of manpower are necessary to support most effectively these strategies? What kinds of individuals and force structures are needed to respond? At what stage can America withdraw, if necessary, from unconventional conflicts with minimum cost? At what stage must conventional forces-in-being be committed to unconventional conflicts? What should determine such commitment? What manpower qualities, quantities, and individual political–psychological dispositions are required for various conflict contingencies?

To undertake a strategic analysis of this type is beyond the scope of this paper and best left to the strategists. Nonetheless, these questions suggest a need to rethink strategic concepts and address new questions, or questions that heretofore have been lost in the maze of strategic nuclear issues or major conventional war considerations.

Manpower–Strategy Linkages

The reorientation of strategic perspectives requires a new relationship between strategists and manpower planners, a closer linkage between strategy and manpower, and a closer manpower "fit" with conflict contingencies. Under conditions of mobilization-type wars and major conventional wars, where conscription and the use of reserves are distinct possibilities, a particular set of manpower policy issues is relevant. In other scenarios, manpower planners should be guided by the specific needs of particular types of forces, for example, those whose missions are aimed at special operations and unconventional conflicts. These may require a different set of manpower questions ranging from recruiting and retention incentives to a focus on particular kinds of individuals most likely to succeed in certain combat situations. A decreasing recruiting base coupled with little possibility of conscription may necessitate a set of strategies reflecting reduced troop levels. When troop levels are at such a premium, it is especially important to fashion manpower requirements aimed direcly at the most likely conflict contingencies.

For example, it is likely that units organized to carry out special operations (as defined here) are best served by individuals who are particularly oriented to the fighting ethic—aggressive, outgoing, and team spirited. In the special forces, where missions are low-intensity conflicts (as defined here), individuals should have a high tolerance for frustration, patience, sensitivity to other cultures, adaptability, communication skills, and the ability to function without reference to the psychological support provided by an American environment.

In sum, having better qualified and psychologically suited individuals in smaller numbers may be a more effective option than maintaining existing levels of more generally qualified individuals. This correlates closely with a strategic orientation that focuses on the conflicts most likely to occur in the immediate period.

SUMMARY

Since the end of the Vietnam War, executive-congressional relationships, the attitudes of major political actors, and the opinions of the American people have become particularly important in determining the role of the military in foreign and national security policy. The commitment of American forces beyond existing treaty obligations is likely to generate reaction from within the American political system. It is common knowledge that Congress is highly sensitive to such issues as covert operations and the foreign involvement of American forces (Grenada notwithstanding). It is presumed by many that any commitment of American forces will open the gates to another Vietnam. Additionally, there are many prominent in American politics who oppose any military force commitment as contrary to democratic principles. Finally, the American way of war, coupled with the American tendency to adopt a Pearl Harbor mentality, makes it extremely difficult to generate realistic perceptions of small wars and low-intensity types of conflicts. This has its corollary in terms of manpower issues in the sense that the volunteer military system is perceived by some to have separated itself from the mainstream of society.

At the end of 1970, a considerable body of opinion viewed the military as a last-resort employer. Concerns expressed publicly by military spokesmen gave the impression that many recruits were in categories III and IV and were high school dropouts. This was also a period in which resources available to the military were deemed inadequate by many military authorities. These developments, compounded by the "fall out" from Vietnam and the post-Vietnam domestic political environment, resulted in an increasingly widespread perception that U.S. military capability had declined markedly. It was during this period that a senior army officer charged that the American army was a "hollow army." At the same time, manpower considerations seemed to reinforce, even if only indirectly, the nation's hypercautious view regarding the commitment of military forces outside existing treaty obligations.

There are similarities in the contemporary period. The fears of any military involvement perceived to be beyond existing capabilities generates fears of conscription and the consequent heavy cost to middle-class Americans. One commentator has gone so far as to call such attitudes military illiteracy. Reviewing the U.S. military record of the last fifteen years and labeling it "abominable," the commentator states that " ... if large segments of the American people have come to doubt both the desirability and the efficacy of the use of American forces, such doubt must seem inevitable and reasonable."[33] Although other recent commentators point to a resurgence of U.S. military spirit, there remains a strong current of opinion that is skeptical about the use of military force and, indeed, the capacity of the military instrument. In this context, the resurgence of a military spirit may be quite distinct from the use of military force in overseas ventures.

In sum, the Vietnam fall out and fears that any American military commitment could lead to mobilization-type wars, with the attendant need for manpower beyond the volunteer military, have created an extremely cautious attitude on the part of many Americans regarding foreign involvement. At the same time, American strategy and manpower planning are driven by a traditional strategic orientation that may well be unrealistic. The most likely future conflicts will be unconventional, conflicts with little conceptual coherence or strategic formulation. Manpower planners, therefore, are functioning in a policy-strategy vacuum in virtually all contingencies except those at the extreme ends of the conflict spectrum.

Under conditions of increasing manpower constraints in a continuing volunteer environment, there is a need to link strategic and manpower considerations more closely. Although some attention is being given to this issue, strategic scenarios continue to fail to give serious attention to the most likely conflicts. Furthermore, while manpower planners focus primarily on manpower levels and population cohorts, strategists stress the nature and character of major conflicts, both nuclear and conventional, giving little attention to the impact of manpower considerations on the implementation of various strategic options. Strategists and manpower planners need to develop a common set of strategic concepts and manpower premises, with the idea that the winner in most contemporary conflicts may not be the one who gets there first with the most, but the one who gets there quickly with the best.

NOTES

1. Max Hastings, *Overlord: D-Day and the Battle for Normandy* (New York: Simon and Schuster, 1984), p. 317.
2. Ibid., p. 151.
3. Ibid., p. 50.
4. Ibid.
5. Harry G. Summers, Jr. *On Strategy: The Vietnam War in Context* (Carlisle Barracks, Penn.: U.S. Army War College, 1981), p. 25.
6. For a discussion of professionalism in the context of the deterrence and battle orientation see Morris Janowitz, *The Professional Soldier: A Social and Political Portrait* (New York: The Free Press, 1971), especially the prologue and chapter 20. See also Morris Janowitz, "Beyond Deterrence: Alternative Conceptual Dimensions," in *The Limits of Military Intervention*, ed. Ellen Stern (Beverly Hills, Calif.: Sage, 1977), pp. 369-89.
7. Jonathan Alford, "Deterrence and Disuse: Some Thoughts on the Problem of Maintaining Volunteer Forces," *Armed Forces and Society* (Winter 1980): 247-56.
8. Charles C. Moskos, "From Institution to Occupation: Trends in Military Organization," *Armed Forces and Society* (Fall 1979): 41-50.
9. Carl Von Clausewitz, *On War*, ed. Anatol Rapoport (Baltimore, Md.: Penguin Books, 1968), p. 241.
10. Bruce Palmer, Jr., "Strategic Guidelines for the United States in the 1980s," in *Grand Strategy for the 1980s*, ed. Bruce Palmer, Jr. (Washington, D.C.: American Enterprises Institute, 1978), p. 73.
11. Michael Howard, *The Causes of War and Other Essays*, 2d ed., rev. (Cambridge, Mass.: Harvard University Press, 1984), p. 112.
12. Russell F. Weigley, *The American Way of War: A History of United States Military Strategy and Policy* (Bloomington, Ind.: University Press, 1977), p. 477.
13. The Organization of the Joint Chiefs of Staff, *Military Posture FY 1985* (Washington, D.C.: U.S. Government Printing Office, 1984), p. 8.
14. See Clinton H. Schemmer, "House Panel Formed to Oversee Special Ops Forces," *Armed Forces Journal International* (October 1984): 15. Interestingly, there is virtually no mention of special operations forces in *U.S.*

Defense Policy, 3d ed. (Washington, D.C.: Congressional Quarterly Inc., 1983) or in Asa A. Clark IV, Peter W. Chiarelli, Jeffrey S. McKitrick, and James W. Reed, eds., *The Defense Reform Debate: Issues and Analysis* (Baltimore, Md.: The Johns Hopkins University Press, 1984).

15. Clark et al., *The Defense Reform Debate*, especially chapter 4. Interestingly, the creation of light divisions in the American army was temporarily hampered by attempts to fashion these units to respond to all types of contingencies. For example, plans were made to assign additional units and increase the division's manpower; hence, its logistical and mobility requirements also were increased, thus shaping it into an almost standard division configuration. Only a directive by the army chief of staff forestalled such developments.
16. Secretary of Defense Weinberger's statement on the criteria for using military force is a case in point. If strictly interpreted, such criteria would preclude the use of the military in all but the clearest cases and only in response to clear threats to American security. See Walter Andrews, "U.S. Lists 6 Criteria for using Troops," *The Washington Times*, 30 November 1984, p. 1.
17. See, for example, Franklin D. Margiotta, James Brown, and Michael J. Collins, eds., *Changing U.S. Military Manpower Realities* (Boulder, Colo.: Westview Press, Inc., 1983); John B. Keeley, ed., *The All-Volunteer Force and American Society* (Charlottesville: University Press of Virginia, 1978); William J. Taylor, Jr., Eric T. Olson, and Richard A. Schrader, eds., *Defense Manpower Planning: Issues for the 1980s* (New York: Pergamon Press, Inc., 1981); and Martin Binkin, *America's Volunteer Military: Progress and Prospects* (Washington, D.C.: The Brookings Institution, 1984).
18. See, for example, Binkin, *America's Volunteer Military*, pp. 41-63 and Clark et al., *The Defense Reform Debate*, pp. 123-76. Part of this debate includes the proper role of reserve forces, not only from the perspective of declining manpower in the active forces but also from the vantage point of strategic concerns regarding the U.S. armed forces' ability to respond to major conflicts and budgetary constraints.
19. William J. Taylor, Jr., and Paul R. Ingholt, "Manpower Issues for the 1980s," in *Strategic Requirements for the Army to the Year 2000*, ed. Robert H. Kupperman and William J. Taylor, Jr. (Lexington, Mass.: D.C. Heath, 1984), p. 143.
20. Ibid., p. 144.
21. Morris Janowitz, *The Reconstruction of Patriotism: Education for Civic Consciousness* (Chicago: The University of Chicago Press, 1983); Stephen D. Wesbrook, *Political Training in the United States Army: Reconsideration*, Mershon Center Position Papers in the Policy Sciences, no. 3 (Columbus, Ohio: Mershon Center, March, 1979); and William Darryl Henderson, *Cohesion: The Human Element in Combat* (Washington, D.C.: National

Defense University Press, 1985). See also Sam C. Sarkesian, ed., *Combat Effectiveness: Cohesion, Stress, and the Volunteer Military* (Beverly Hills, Calif.: Sage Publications, 1980) and the Defense Management Study Group on Military Cohesion, *Cohesion in the U.S. Military* (Washington, D.C.: National Defense University Press, 1984).

22. Alan Ned Sabrosky, *Defense Manpower Policy: A Critical Reappraisal*, Monograph no. 22 (Philadelphia: Foreign Policy Research Institute, 1978), p. 65.
23. Gregory D. Foster, "The Effect of Deterrence on the Fighting Ethic," *Armed Forces and Society* 10, no. 2 (Winter 1984): 290.
24. For a detailed explanation of these concepts see Sam C. Sarkesian, "Low-Intensity Conflict: Concepts, Principles, and Policy Guidelines," *Air University Review* 46, no. 2 (January–February 1985): 4–23.
25. Howard, *The Causes of War*, p. 112.
26. Sabrosky, *Defense Manpower Policy*, p. 65. See also Charles Doe, "Analyst Warns U.S. Forces Are Over-Committed," *Army Times*, 8 October 1984, pp. 23 and 65.
27. Summers, *On Strategy*, p. 7.
28. Ibid., p. 25.
29. Howard, *The Causes of War*, p. 22.
30. Ibid., p. 105.
31. Ibid., pp. 109–110.
32. Sun Tzu, *The Art of War*, trans. and intro. Samuel B. Griffith (New York: Oxford University Press, 1971), pp. 64 and 77.
33. For example, see the exchange between Jeffrey Record and Vincent Davis in *SAIS Review* 5, no. 1 (Winter-Spring 1985): 270–274.
34. Philip Gold, "Force and Military Illiteracy," *Chicago Tribune*, 4 February 1984, section 1, p. 9.

IV MILITARY MANPOWER OPTIONS

6 MANPOWER PROCUREMENT OPTIONS
The Influence of Demography, Technology, and Budgets

Martin Binkin

The early 1980s was a period of remarkable success in terms of staffing the nation's armed forces with qualified volunteers. The services have enjoyed both a bountiful harvest of high-caliber recruits and a substantial improvement in the proportion of troops seeking a military career.

This record is all the more impressive when contrasted with the immediately preceding period, during which the qualifications of military recruits had dropped to postwar lows, the services were having trouble keeping skilled specialists and technicians, and doubts about the effectiveness of U.S. military forces were reinforced by unsettling events in Southwest Asia. By the end of the 1970s, many legislators and policymakers questioned the viability of raising armed forces by strictly voluntary means.

Despite this dramatic turnaround, some observers remain skeptical. Critics contend that the current successes have been achieved at the expense of quantity, maintaining that today's armed forces have been sized to fit the military's ability to recruit volunteers and are too small to meet national security commitments. Others worry that the abolition of conscription and the adoption of the total force con-

---

This chapter has been adapted from Martin Binkin, *America's Volunteer Military: Progress and Prospects* (Washington, D.C.: The Brookings Institution, 1984) and Martin Binkin, "Manpower" in *American Defense Annual, 1985-1986*, ed. George E. Hudson and Joseph J. Kruzel (Lexington, Mass.: Lexington Books, 1985).

cept have necessitated an excessive—and risky—reliance on reserve forces, which are ill equipped and, in some instances, undermanned for assigned tasks. Finally, the same ideological concern that has long bothered critics of the volunteer concept persists: a weakening of the traditional belief that each citizen has a moral responsibility to serve his country. These concerns, however, have not yet been taken seriously enough to dampen the euphoria of recent successes, much less to prompt serious consideration of changes to the method of military manpower procurement.

The greater threat to voluntary recruitment lies in the future as several trends converge within the next decade or so, possibly creating an adverse supply and demand situation. On the supply side, the pool of prospective recruits will shrink as the baby boom generation passes beyond military age; additionally, if the economy continues to recover, fewer among them are likely to be interested in serving. On the demand side, anticipated increases in the size of the armed forces and expected advances in weapons technology could increase the military's need for recruits and perhaps for better ones as well. If these events occur, the armed forces will be squeezed between a diminished supply of prospective high-quality volunteers and an increased demand for better recruits, thus driving up the price of military manpower. It is unlikely, moreover, that deficit pressures will allow growth in defense outlays sufficient to absorb the large bills expected as a result of the recent surge in defense investment and to cover substantial increases in personnel costs as well.

The severity of the problem will depend on the extent to which additional people are needed by the services to man the expanded force structure, on the one hand, and better qualified people are needed to man more sophisticated weaponry, on the other. If these demands are large, the practical options narrow and policymakers ultimately could be forced to choose between smaller forces, simpler systems, or a return to conscription.

LOOKING BACK

The vitality of the nation's all-volunteer armed forces has been a subject of continuous debate since the end of conscription in 1973. In the absence of agreed upon measures of output as a means of assessing the voluntary concept, the most widely used gauge has been the "quality" of recruits; that is, their educational attainment and entry test scores.

Table 6-1 shows the trends in recruit quality during the draft and volunteer eras. For the first several years following the end of conscription, the armed services as a whole attracted roughly the same proportion of high aptitude (categories I and II) recruits as they had during the draft years, a larger proportion with average aptitude and a smaller percentage in the lower acceptable category. This successful transition can be ascribed largely to the substantial increases in pay instituted in 1972 to launch the volunteer experiment and to a lesser extent to the slackened labor market in 1973-74.

In sharp contrast to the 1974-76 experience, the late 1970s was a difficult period for the volunteer force as recruit quality plummeted. Of all volunteers who entered the military services between 1977 and 1980, for example, 28 percent were in the lowest acceptable category. The army was the most seriously affected; as Table 6-2 shows, 44 percent of recruits scored in the bottom category. Several factors accounted for the turnaround: (1) military pay fell behind private sector pay as the Ford and Carter administrations imposed a series of pay caps; (2) the relative improvement in the nation's economy in the late 1970s was accompanied by declining rates of unemployment; and (3) the Vietnam-era GI bill was replaced by a less generous plan while civilian student aid programs were expanded.

It should be noted that, because of confusion about entry test scores, the armed services apparently did not realize the extent of their problems until 1980, when it was revealed that the test scores of a large number of recruits had been overstated; many recruits who would otherwise have been ineligible were accepted by the armed services. The magnitude of the error was substantial; for example, in contrast to the original belief that only 9 percent of the recruits who entered the army in 1979 had scored in category IV, corrected scores placed close to half in that category.

Alarmed by these developments, Congress enacted substantial increases in pay and benefits in 1980 and 1981 to shore up inadequate military wages and spur recruitment of higher quality youth. This was accompanied by a deepening recession and further deterioration of the civilian employment prospects for American youth. The army, for its part, intensified its recruitment efforts by expanding its advertising program and enlarging its corps of recruiters. Furthermore, an improved educational benefits package was made avail-

Table 6-1. Percentage Distribution of Recruits, by Aptitude Category and Level of Education, All Services, Selected Fiscal Years, 1960–1984.

Aptitude Category and Educational Level	Draft Era			Volunteer Era		
	1960–64	1965–69	1970–73	1974–76[a]	1977–80[b]	1981–84
Armed Forces Qualification Test Category						
I and II (above average)	38	38	35	35	29	38
III (average)	49	41	45	55	43	50
IV (below average)	14	21	22	10	28	12
Education Level						
College degree	2	3	5	1	1	2
Some college	11	15	13	5	4	7
High school diploma	51	56	52	60	66	79
Total (having at least high school diploma)	64	74	70	66	71	88

a. Includes the quarter July–September 1976 to account for the transition to the fiscal year beginning October 1976.
b. Aptitude category data reflect corrections made to misnormed scores as described in text.
Source: Martin Binkin, *America's Volunteer Military: Progress and Prospects* (Washington, D.C.: The Brookings Institution, 1984), table 2, p. 8.

Table 6-2. Percentage Distribution of Army Recruits, by Aptitude Category and Level of Education, Selected Fiscal Years, 1960–1984.

Aptitude Category and Educational Level	Draft Era				Volunteer Era		
	1960	1964	1969	1972	1974–76[a]	1977–80[b]	1981–84
Armed Forces Qualification Test Category							
I and II (above average)	32	34	35	33	29	18	33
III (average)	51	47	38	49	54	37	50
IV (below average)	17	19	27	18	17	44	17
Educational Level							
College degree	5	5	6	4	1	1	2
Some college	26	15	19	11	4	3	7
High school diploma	36	50	45	46	50	58	78
Total (having at least high school diploma)	67	70	70	61	55	62	87

a. Includes the quarter July–September 1976 to account for the transition to the fiscal year beginning October 1976.
b. Aptitude category data reflect corrections made to misnormed scores as described in text.
Source: Binkin, *America's Volunteer Military*, table 3, p. 9.

able to selected army enlistees at the same time that the Reagan administration was threatening reductions in student aid programs.

The separate effects of these changes are difficult to measure, but together they worked, as the qualitative characteristics of recruits improved markedly beginning in 1981. Since then, the services have recruited unprecedented proportions of high school graduates and effected a sharp improvement in the aptitude profile. The turnaround was especially conspicuous in the army, which had experienced the greatest difficulties in the late 1970s (see Table 6-2).

But just as substantial pay increases and a faltering economy contributed to a dramatic improvement in recruitment in the early 1980s, the effects of meager pay raises since 1981 and a recovering economy have taken their toll. Three successive years of 4 percent pay increases diminished the ratio of military to civilian pay, especially at the entry level, while the nation's unemployment rate between 1982 and 1984 was dropping from 9.7 percent to 7.5 percent.

The first signs of a downturn in recruitment appeared in fiscal 1984 as the army reported a decline in the number of new enlistment contracts negotiated in that fiscal year. At the beginning of fiscal 1985, in fact, the army's delayed entry pool (those signed up in anticipation of entering within twelve months) was about 8,000 short of its overall goal and 6,000 short of its goal for high-quality males. Moreover, the army, navy, and air force fell somewhat short of their contract objectives for the first quarter of fiscal 1985.

Meanwhile, the initial results of the 1984 Youth Attitude Tracking Study, which measured the propensity of youth to enlist in the armed forces, provided another ominous sign. Compared with im-

Table 6-3. Percentage of Males, Ages 16 to 21, with a Positive Propensity for Military Service, by Service of Choice, 1979-1984.

Service	1979	1980	1981	1982	1983	1984
Any service	30.0	33.7	34.3	35.8	35.4	29.9
Army	12.9	14.6	15.0	16.0	17.5	14.3
Navy	14.5	14.4	15.4	14.4	13.0	10.9
Air force	16.6	20.6	20.9	18.7	18.8	15.3
Marine corps	10.8	12.3	12.4	11.7	12.1	9.6

Source: *Air Force Times*, 25 March 1985, p. 8.

mediately preceding years, a substantially lower proportion of young males indicated an intention to join the military (see Table 6-3). The figures for the air force, navy, and marines were the lowest recorded since the annual surveys began in 1975, while the army fell to below 1980 levels.

THE NEXT DECADE

These developments, in themselves, are not of major consequence, but the prospect that they may usher in a challenging period for the all-volunteer force should be of concern. Keeping the military staffed with qualified volunteers over the next decade could become progressively more difficult as demographic, technological, and economic factors converge.

Demographics

The armed forces will confront a declining supply of prospective volunteers over the next decade as the nation's youth population dwindles. In contrast to 1985, when close to 3.7 million Americans turned eighteen, only 3.2 million are expected to reach that age in 1995 (see Table 6-4). This 13 percent decline in the age group from which the military traditionally seeks its volunteers will make the recruitment task more difficult. A rough indication of the magnitude of the challenge is provided by calculating the proportion of qualified and available males who, in the long run, would have to volunteer for military service before reaching age twenty-three if the active and reserve forces are to meet their projected recruiting needs.[1] This can be done by following one age group through time—excluding those who, based on past experience, are *not* likely to volunteer (third- and fourth-year college students) and those who *cannot* volunteer (because they are mentally, physically, or morally unqualified).

As Table 6-5 shows, during the 1984-88 period, an average of 1.8 million noninstitutionalized males will turn eighteen each year. If history is a guide, however, 525,000 will enter college and another 526,000 will not meet minimum standards for entry into the armed forces. Thus, during this period the pool of prospective volunteers

Table 6-4. Projected U.S. Population Aged 18 to 21, by Sex and Race, Selected Years, 1981–1995, in Thousands.

Category	1981	1983	1985	1987	1989	1991	1993	1995
Male	8,618	8,356	7,821	7,356	7,404	7,197	6,702	6,608
White	7,282	7,010	6,509	6,085	6,098	5,864	5,405	5,331
Black	1,147	1,145	1,102	1,053	1,070	1,071	1,022	994
Other	190	201	210	218	236	262	275	283
Female	8,401	8,142	7,621	7,164	7,197	6,984	6,495	6,386
White	7,059	6,799	6,312	5,896	5,897	5,666	5,220	5,137
Black	1,168	1,161	1,116	1,067	1,081	1,076	1,022	990
Other	174	182	193	201	219	242	253	259
Total	17,019	16,498	15,442	14,520	14,601	14,181	13,197	12,994

Source: U.S. Department of Commerce, Bureau of the Census, *Projections of the Population of the United States: 1977 to 2050*, Series P-25, no. 704 (Washington, D.C.: Government Printing Office, 1977), pp. 40–60. Figures are rounded.

Table 6-5. Proportion of Qualified and Available Males Required for Military Service, Selected Periods, 1984-1985 (*in thousands unless otherwise indicated*).

		Annual Average		
		1991-95		
Category	1984-88	A	B	C
Total Noninstitutionalized Eighteen-year-old Males	1,800	1,612	1,612	1,612
Minus: college enrollees less first- or second-year dropouts	525	464	464	464
Minus: unqualified	526	461	461	617
mental	337	291	291	486
physical or moral	189	170	170	131
Equals: Qualified and Available Male Pool	749	687	687	531
Total Male Recruit Requirements	376	376	410	410
Active forces	278	278	287	287
Reserve forces	98	98	113	113
Percent of Pool Required	50	55	60	77

Source: Derived from Martin Binkin, *America's Volunteer Military: Progress and Prospects* (Washington, D.C.: The Brookings Institution, 1984), pp. 31-39.

would be exhausted at about 750,000 a year, out of which the armed forces would need to attract 376,000, or about half of them, to meet planned recruit requirements. By the early 1990s, the figure grows to 55 percent as the noninstitutionalized eighteen-year-old cohort drops to about 1.6 million (see column A). Not factored into this calculation, however, are the effects of the anticipated increase in the size of the armed forces and the possible adoption of more stringent entrance standards to meet the demands of a more sophisticated force.

The Prospective Military Buildup

In the face of budgetary pressures, the Reagan administration has apparently scrapped its earlier blueprint to expand substantially the size of the armed forces commensurate with its military buildup. Modest increases are still planned for the navy, which projects an increase of 43,000 by 1990, largely to man the 600-ship fleet. The air force plans to add 33,000 by the end of the decade to support the deployment of cruise missiles to Europe, the expansion of tactical fighter units, and the fielding of the MX. The army, which abandoned earlier plans to expand its active component, intends to increase its reserve component by over 110,000 by 1990.

Adding 260,000 to the active and reserve military rolls could be expected to increase the average annual requirements for male recruits by about 34,000, raising the total to 410,000, or 60 percent of the qualified and available population in the early 1990s (column B in Table 6-5).[2]

Technology and Qualitative Standards

In addition to the prospect that the armed services may have to attract more volunteers, the possibility exists that they may have to seek highly qualified volunteers capable of absorbing more complex technical training. As the armed forces field increasingly sophisticated weapon systems, their requirements for specialists and technicians are expected to grow.

The influence of technology on the military's skill mix in the postwar period is traced in Table 6-6, which shows the conspicuous shift from work requiring general military skills toward tasks requiring special expertise. The sharpest changes have taken place in technical skills (those involving computer specialists, electronics technicians, medical technicians, and the like), the service and supply occupations, and general combat skills. The proportion of technical jobs, which has always been higher in the equipment-intensive navy and air force, has increased markedly in the army and, to a lesser extent, in the marines. For the army, the increase in technical specialization has been at the expense of service and supply occupations and combat skills.

These trends are likely to continue into the future as new generations of weapon systems enter the inventory. As a result of the Reagan administration's program to bolster U.S. defense capabilities, the military services are undergoing the most rapid, comprehensive, and simultaneous peacetime modernization in the nation's history.

Table 6-6. Distribution of Trained Military Enlisted Personnel by Major Occupational Category, Selected Years.

Major Occupational Category	Percent of Trained Enlisted Personnel				
	1945	1957	1963	1973	1984
White Collar	28	40	42	46	47
Technical workers	13	21	22	25	30
Electronics	(6)	(13)	(14)	(18)	(22)
Other	(7)	(8)	(8)	(7)	(8)
Clerical workers	15	19	20	20	18
Blue Collar	72	60	58	54	53
Craftsmen	29	32	32	28	27
Service and supply workers	17	13	12	13	11
Infantry, gun crews, and seamanship specialists	24	15	14	13	16

The Reagan program emphasizes improvements in: (1) the survivability of U.S. forces; (2) the detection and tracking of enemy forces; (3) the accuracy and lethality of U.S. weapons; and (4) battle management capabilities, both nuclear and conventional.

Advances in aerodynamics and propulsion can be expected to accompany developments in materials, chemistry, and structures, but these are likely to pale in comparison to the effects of emerging electronics technologies. Indeed, many knowledgeable observers believe that the U.S. armed forces are on the threshold of a major breakthrough in electronics that, like nuclear weapons and jet propulsion of earlier eras, will revolutionize warfare. The individual military services will be affected in different ways.

The army, for example, with its growing emphasis on offensive-minded, deep strike doctrine, is developing operational concepts that will make necessary the integration of C^3 (command, control, and communications), active and passive surveillance, real-time tactical information dissemination, targeting, and weapons delivery employing a wide stand-off area or point suppression guided by precise navigation or active target designators.

The technological key to these capabilities, according to one expert observer, is the computer chip. "Exploiting the chips," wrote one senior army commander, "will allow us to achieve surprise, gain and maintain momentum, seize the initiative, cause the enemy to react, and set the stage for the confusion and paralysis of the enemy."[3] Advances in electronics are expected to extend tactical intelligence capabilities by permitting the army to see over extended ranges at night and under any weather conditions, to detect and track enemy formations, to locate the enemy's electronic emissions and artillery batteries, and to listen to the enemy's communications, all while blinding the enemy's intelligence sensors. To translate, classify, and analyze the vast quantity of information that will be generated, electronically rich command, control, communications, and intelligence systems are to be fielded. In contrast to the information distribution systems currently in place, dependent for the most part on manual operations, the next-generation system is expected to be more fully automated and, by exploiting such advances as distributive computer processing, possess the ability to coordinate input from various sensors and resolve ambiguities or errors.

The air force likewise anticipates that the electronics revolution will have dramatic effects on the future application of airpower. Microminiaturization is expected to lead to the total integration of aircraft avionics, the transition from "smart" to "brilliant" weapons, and with developments in artificial intelligence and other software advances, usher in a golden age of C^3I.

Microminiaturization is also expected to overcome the problems of current signal processors, which are not fast enough to collect and interpret data from the various types of exotic sensors (infrared and millimeter-wave radar) under development. Finally, improved antennas, millimeter waves, advanced signal processing, and radiation hardening are expected to provide an ability to gather information in hostile environments without appreciable constraints.

For the navy, an ability to install potent, long-range weapon systems in smaller spaces is expected to be among the important payoffs of the new electronics technologies. This would give the navy the choice of installing more systems for a given ship size or of constructing smaller (and probably less detectable) ships for a given capability. The new technologies will also have important implications for C^3I, not so much in providing an integrated and automated system (which, unlike the army, the navy is already on its way to achieving), but to permit coverage broad enough to include ocean-sized areas and ultimately tactical integration of ocean areas.

Finally, the marines, whose ground component has traditionally eschewed high technology, also plan to exploit electronic developments. The existing amphibious assault vehicles (LVTP-7) are being upgraded with all-electric weapon stations and the systems on its successor (LVT-X) are likely to be even more dependent on electronics. Also a family of "smart" munitions, made possible by the technological developments discussed previously, will be deployed with units of marines in the 1990s.

The impact of these developments on the military's requirements for skilled personnel is difficult to pin down and, in fact, a subject of some controversy. One school of thought contends that emerging electronic technologies will render new systems easier to operate and maintain and thus reduce the need for highly qualified personnel. The "exploding technology of microprocessors in automatic test equipment, built-in-test, and in simulators and training devices," according to then Under Secretary of Defense for Research and Engineering William J. Perry, will "enable our forces to effectively and efficiently operate and maintain their equipment."[4]

Perry's position rests on the supposition that technology will be exploited to simplify the operations and maintenance of weapon systems. Historically, however, this has been the exception rather than the rule as "most often technical advance is used to stretch the performance of defense systems to the limit."[5] A study of the impact of technology on naval aircraft systems concluded that "although new technology has improved component reliability (failures per part per flight hour), it has also permitted an increase in density of functions and capabilities (number of parts per system). This has resulted in overall decreases in system reliability and increases in maintenance manpower requirements."[6]

More recent data indicate that the newer systems being fielded will be easier to operate. Table 6-7 shows the results of on-the-move range firing by 1,131 crews equipped with M-60 and M-1 series tanks. M-1 tank crews consistently scored more tank "kills" than M-60 crews with similar aptitudes. More interestingly, the number of tank kills by M-1 crews was less influenced by their aptitudes.[7] These results suggest that the more complex M-1 system is easier to employ than its predecessor and, therefore, capable of being operated by lesser qualified personnel for a given level of effectiveness. But whether this is also true for maintenance remains to be seen. The numbers and qualifications of maintenance personnel will de-

Table 6-7. Tank Crew Performance.

AFQT Category Gunner/Tank CMDR	Tank Kills		Percent Improvement
	M60	M1	
I	10.23	12.75	25
II	9.51	12.47	31
IIIA	8.52	12.05	41
IIIB	7.47	11.57	55
IV	5.84	10.72	84

Source: Barry L. Scribner et al., "Are Smart Tankers Better Tankers: AFQT and Military Productivity" (West Point, N.Y.: Office of Economic and Manpower Analysis, U.S. Military Academy, 1984).

pend on the reliability of the equipment and the maintenance workload, for which few data are publicly available.

Although lacking rigorous foundations, the military services are projecting increasing requirements for skilled specialists and technicians. The additional sailors needed to man the expanded fleet will have to meet stricter standards as the navy's occupational mix adapts to high technology. Between 1982 and 1987, for example, semitechnical billets are expected to increase by about 6 percent, technical jobs by 10 percent, and highly technical positions by close to 13 percent.[8] It is also estimated that the air force's need for people with high aptitudes for electronics skills will increase by one-third by the year 2000.[9]

Emerging technology is expected to alter the occupational mix of the army as it moves from the electromechanical to the electronic age. Since many of the systems now under development are heavily dependent on electronic data processors, the inner balance of the army will shift in the direction of skills related to communication and electronics, military intelligence, and ordnance and away from infantry, armor, and air defense. Given the present course, by 1990 the signal corps will replace the infantry as the largest army branch.[10]

Partly in anticipation of the shift toward a more skilled force, the army has already adopted highly ambitious recruitment goals for fiscal 1986 and beyond: at least 90 percent high school graduates, at least 65 percent with enlistment test scores that rank in the top half of the population, and not more than 10 percent in the lowest test score category.

To the extent, then, that advances in technology lead to a richer mix of technical skills, the recruitment task will become more difficult since the pool of qualified youth can be expected to shrink. As Table 6-8 illustrates, given today's minimum standards, a much smaller proportion of American youths can qualify for entry into higher skill jobs (especially in electronics) as compared to other occupations.

The changes that will be required in the quality mix are difficult to predict, but the potential effects can be illustrated. At the extreme, if the army were to impose the more demanding entry standards now used by the air force, the pool of qualified and available male youths would shrink by over 20 percent and three out of every four eventually would have to enlist to meet projected military requirements in the early 1990s (column C in Table 6-5).[11]

Economic Recovery

While it can be concluded, based on the above discussion, that the recruitment task is bound to become more difficult over the next decade, the extent to which the armed forces will be able to meet their goals will depend to a large degree on economic factors.

The military services have traditionally experienced more difficulty in attracting higher quality volunteers during periods of economic expansion and declining rates of unemployment (for example, 1976-79). Alternatively, when the labor market slackens, as it did in 1973-74 and again in 1980-83, the number and quality of applicants increases.

Thus, it is to be expected that economic recovery would once again create a more competitive environment for the armed forces. In fact, the decline in enlistment contracts in fiscal 1984, as discussed above, could well be tied to the drop in the unemployment rate that began in fiscal 1983. The recovery is expected to continue for the remainder of the decade, with the administration projecting a continued decline in the unemployment rate from 7.0 percent in 1985 to 5.8 percent in 1990.[12] A drop of this magnitude would be expected to result in a decline of up to 10 percent in high-quality,

Table 6-8. Percentage of American Youths (18-23 Years) Who Would Qualify for Selected Occupations in the Army and Navy (Arranged According to Occupational Area), by Sex and Racial/Ethnic Group.

Sex and Racial/ Ethnic Group	Mechanical Occupations	
	Army: Heavy Construction Equipment Operator	Navy: Machinist's Mate
Male		
White	82.2	76.0
Black	29.8	20.5
Hispanic	47.4	38.5
Total	73.1	66.3
Female		
White	71.3	51.9
Black	21.0	12.6
Hispanic	26.3	17.7
Total	61.6	44.4
	Administrative Occupations	
	Army: Administrative Specialist	Navy: Yeoman
Male		
White	63.6	35.5
Black	19.4	6.5
Hispanic	34.4	14.6
Total	56.0	30.4
Female		
White	74.2	48.9
Black	27.3	10.9
Hispanic	37.1	15.7
Total	65.4	41.6

Table 6-8. continued

	General Occupations	
Sex and Racial/ Ethnic Group	Army: Military Police	Navy: Mess Management Specialist
Male		
White	78.3	78.0
Black	27.4	30.1
Hispanic	44.3	44.8
Total	69.5	69.7
Female		
White	77.7	79.1
Black	29.1	32.0
Hispanic	37.6	37.7
Total	68.5	70.0
	Electronics Occupations	
	Army: Satellite Communi- cations Equipment Repairman	Navy: Electronic's Technician
Male		
White	29.7	50.7
Black	3.4	8.4
Hispanic	9.2	21.4
Total	24.9	43.3
Female		
White	11.7	30.4
Black	0.4	3.4
Hispanic	2.2	7.0
Total	9.5	25.2

Source: U.S. Department of the Navy, *Enlisted Transfer Manual*, NAVPERS 15909C (Washington, D.C.: U.S. Department of the Navy), pp. 7-11, 7-12; U.S. Department of the Army, *U.S. Army Formal School Catalog*, Pamphlet 351-4 (Washington, D.C.: U.S. Department of the Army), 1 April 1983; "Profile of American Youth" (Alexandria, Va.: Defense Manpower Data Center).

male volunteers by the end of the decade. If the less-optimistic projections of the Congressional Budget Office are used, the unemployment rate will drop to 6.3 percent by 1990, in which case the decline in high-quality, male volunteers would be in the neighborhood of 5 to 6 percent.[13]

OPTIONS

Despite the successes of the early 1980s, the ability to man the nation's armed forces with enough qualified volunteers cannot be taken for granted, given the uncertainties involved. As a matter of fact, if the volunteer system is to survive the challenges of the next decade, as described above, additional investment in the military payroll and recruiting budget may be necessary. Given the budgetary outlook, however, options for reducing the demand for male volunteers or expanding the pool of qualified prospects will become more attractive. In any event, advocates of conscription, anticipating the futility and, indeed, the risk of attempting to maintain a manpower procurement system so sensitive to the vagaries of the marketplace and the budgetary process, will continue to urge the nation to prepare for a return to some form of compulsory service.

Compensation and Recruitment

Daunting though the prospect might appear for attracting an increasing proportion of the male youth population, a number of informed observers are optimistic that the armed forces will be able to meet their personnel needs for the foreseeable future. The Congressional Budget Office has concluded that if military pay raises keep pace with those in the private sector, all services could be expected to meet congressionally imposed quality constraints through 1988, even though the proportion of male high school graduates would probably fall below the record-setting levels of the early 1980s.[14] Other manpower specialists likewise have been optimistic: "Where a few years ago predictions of a return to peacetime conscription by the mid-1980s were common, more prevalent now is the view—which we share—that military strength and quality can be maintained at desired levels so long as the nation maintains its commitment to keeping military compensation, broadly construed, competitive with civilian compensation."[15]

The amount that would have to be invested in the military payroll to attract and retain enough qualified people is unclear. Much would depend on the amount of quality considered necessary and on the state of the economy. According to one estimate, even if military pay raises keep pace with the average change in civilian pay, by 1990 the number of high-quality male enlistees will be close to 30,000 (or about 18 percent) below the 1982 peak level for the entire Defense Department and about 5,000 (or about 10 percent) below the 1982 level for the army.[16] To maintain the high-quality profile of the early 1980s in the face of anticipated declines in the youth population and unemployment rates would require an increase in relative pay of 10 percent, which in fiscal 1985 terms would amount to over $6 billion.[17] Similar results could possibly be achieved at a lower cost if pay increases were applied selectively to skills that are in short supply.

In any event, the nation's commitment to maintaining competitive pay levels appears fragile. Even the strong pro-defense Reagan administration failed to live up to early expectations that pay for members of the armed forces would be a priority issue and, at the least, would keep pace with private-sector pay. Apart from the substantial increase (14.3 percent) granted in fiscal 1982, the administration has capped—and in one instance attempted to freeze—military pay increases at levels below average civilian raises, thereby permitting the ratio of military to civilian pay to fall.

There are few signs, however, that the situation is about to change. In fact, pressures to suppress military pay are likely to increase over the remainder of the decade as, on the one hand, personnel expenditures are squeezed between the large stream of outlays generated by the recent surge in investment, and on the other, growth in total defense spending will probably not exceed 3 percent per year in real terms.

A growing body of opinion holds that additional recruiting resources would be more efficient than pay increases to stave off recruitment problems. Current estimates of recruiter elasticity vary, but it is generally concluded that a 10 percent increase in the number of recruiters would produce at least a 5 percent increase in the number of high-quality enlistees. There is less of a consensus, however, on the range over which this estimate might be appropriate. One analyst has concluded that the point of diminishing returns has not been reached and argues that the upcoming decline in the youth population could be completely offset by a proportionate increase in the number of recruiters.[18]

Manpower Management

In addition to competitive pay levels, the adoption of efficient manpower policies was an important premise of the architects of the volunteer concept. As early as 1973 an assessment of the prospects for success concluded that the all-volunteer force "is likely to prove a feasible proposition, *if* timely measures are taken to reevaluate manpower requirements and standards and to deal with foreseeable recruiting shortfalls."[19] A fresh examination was needed, it was argued, of the role of women; the potential for civilian substitution; the educational, mental, and physical standards for enlistment; and the youth-experience mix in the armed forces—all issues designed to reassess the demand for and supply of *male* volunteers.

Several things have been done since the end of conscription to reduce the requirement for new male recruits: The size of the forces has been modestly cut; the role of women in uniform and of federal civilians has been expanded; and, on average, personnel are remaining in the military for longer periods of time. The net effect has been to reduce average annual male accessions from 364,000 during the first several years of the all-volunteer force to just over 280,000 in the early 1980s. The extent to which the need for male recruits can be trimmed even further by manipulating these policy variables once again needs to be assessed.

Curtailing the Manpower Expansion. Pressure has already been put on the administration by Congress to rethink its plans for expansion by cutting Pentagon requests for additional active military manpower in each of the last three years, while encouraging a larger role for reserve forces. Whether the net effect on the recruitment task would be beneficial, however, would depend on the extent to which a reserve buildup could be accomplished by a greater dependence on signing up veterans leaving active duty rather than by attracting new recruits. If not, the recruitment problem would merely be transferred from the active to the reserve forces, and the overall task would remain essentially unchanged, unless it proves easier to attract high-quality youths to the reserve forces than to the active forces.

More recently, a proposal by Senator Warren Rudman (Republican from New Hampshire) would not only deny the Pentagon's request for additional military manpower for fiscal 1986 but would reduce current strength levels by about 75,000.[20] If enacted, this measure would alleviate pressures on military recruitment, but the overall impact on military effectiveness needs further analyses.

Expanding the Role of Women. The services have also been under continued pressure to widen the opportunities for women. While the expansion over the last decade in the number of military women has been quite dramatic, there remains a widespread feeling that more can be done, especially by the air force, which is least constrained by existing laws and policies regarding combat exposure. In fact, the fiscal 1985 defense authorization bill included a provision requiring that 19 percent of air force recruits in fiscal 1987 and 22 percent in fiscal 1988 be female.

The possibilities for expanding the role of women in services other than the air force are narrower, but the issue deserves further consideration. The nation has yet to come to grips with the basic question of whether the laws and policies that constrain further expansion are justified by valid national security concerns or instead represent vestiges of sexual stereotypes of an earlier era.

Substituting Civilians. The possibility for substituting civilian for military personnel in appropriate jobs also deserves attention. While some progress has been made over the last decade in the utilization of civilians (federal and contract), there is ample evidence that the potential has not been fully exploited. One analysis suggests that the navy could fill at least 17,000 and perhaps as many as 45,000 shore billets with civilians without degrading fleet capabilities.[21] Overall, it has been estimated that there are roughly 300,000 positions now filled by military personnel that could conceivably be filled by civilians, even allowing for overseas deployments and job rotations.[22] Moreover, advances in military technology should open even more opportunities for civilians. With the greater emphasis on "black-box" replacement at the operational unit level, for example, a greater share of the equipment maintenance burden should be absorbed at the civilian-manned depot level.

First-Term–Career Mix. The retention of a larger proportion of military personnel is another means for reducing the requirement for new recruits. This approach, moreover, would also provide an effective corps of specialists and technicians more closely matched to the technological needs of the modern military. The appropriate mix of career and first-term members is a controversial question, influenced by the military's accent on youth and adherence to the traditional pyramidal rank structure. To realize the full potential of a more seasoned military force, however, would require a restructuring of the military compensation system.

Assessing the Quality Mix. Since the recruiting task is so sensitive to the level of manpower quality being sought, it is important that the services' quality requirements be scrutinized. In the face of technological advancements it may well be that soldiers of the future will have to be better qualified. But if, as implied above, the army's quality goals carry too high a price tag, proposed weapons systems should be reexamined with a view toward making them more compatible with the capabilities of the people that the army can realistically expect to attract and retain. This is not to say that the army should eschew high technology; rather it is to suggest that greater emphasis be placed on reliability and maintainability, even if some capability must be sacrificed.

Change the Method of Manpower Procurement

A number of observers are unprepared to accept either the financial costs that are likely to accrue as the price of military manpower rises or the redefinition of military service that has accompanied the volunteer concept. The reinstitution of some form of compulsory service is their recommendation. But many critics of the current situation stop short of proposing a return to conscription that merely meets the needs of the armed forces. Rather, they envision military service as only one aspect of a national service program that would involve a broader spectrum of American youth employed in public service as well.

It is not at all obvious, however, that a return to conscription would vastly improve future prospects. A great deal would depend on the extent to which the army relied on inductees rather than volunteers, which in turn would depend on quality considerations.

Unless the armed forces raised standards and thus reduced the supply of qualified volunteers, say, to meet the demands of emerging technology, relatively few inductees would be needed to make up for recruitment shortfalls. If only small numbers were drafted, the overall caliber of recruits would not improve substantially since draftees would ostensibly be representative of the total qualified population. Moreover, a resumption of conscription of any magnitude, at best, would have little effect on the retention of experienced personnel and, at worst, would diminish the overall experience level of the force since, historically, draftees and draft-motivated volunteers have had a much lower propensity than *true* volunteers to remain in service beyond a first term.

The issue appears moot, at least for the time being, as few legislators appear willing to broach the question. Recently Senator Gary Hart (Democrat from Colorado) and Congressman Robert G. Toricelli (Democrat from New Jersey) co-sponsored a measure that would establish a commission to explore the potential of a system of universal national service, but a similar proposal introduced in 1983 by Representative Leon E. Panetta (Democrat from California) to explore voluntary national service was defeated on the floor, 245 to 179.[23] Practically speaking, however, the little interest Congress has shown in national service peaked in the late 1970s and all but disappeared as the military manpower situation improved.

There is even less congressional interest in reviving the draft. Its principal proponent, Senator Ernest F. Hollings (Democrat from South Carolina) proposed legislation in 1985 to renew conscription while acknowledging the slim chance of passage, at least in the immediate future.[24]

WRAP-UP

What will happen to the military manning situation over the next decade is unclear. The only factor that we can predict with confidence is the decline in the youth population. While the behavior of the economy and the impact of technology are less certain, informed observers agree that the recruitment task is bound to become more challenging over the next decade.

By most economic reckoning, while the number of recruits with high school diplomas and above average aptitude scores can be expected to decline from the unusually high levels of the early 1980s, a deterioration of the magnitude necessary to prompt a serious debate about a return to conscription is difficult to envisage, providing military pay raises keep pace with those in the private sector and economic recovery is not more vigorous than expected.

If, on the other hand, the decline in the youth population is accompanied by a requirement for a substantially larger or more skilled military force, the armed forces could face the prospect of having to attract a sizable proportion of the qualified pool of young men. Under those circumstances, if the problems of the late 1970s are to be avoided, more imaginative manpower policies will have to be pursued.

Practically speaking, in the absence of a serious deterioration in U.S.-Soviet relations, a return to peacetime conscription appears unlikely. If the nation, then, is to rely strictly on volunteers for the foreseeable future, every effort should be made to ensure that they are up to the task of protecting U.S. national security interests. Administration officials and legislators should abandon the boom-or-bust philosophy of military pay increases in favor of stable, predictable pay policies. Military leaders should be more willing to entertain new ways of managing their manpower resources, even if it means breaking with tradition. Finally, human factors engineers should have a larger role in the research and development process to ensure that weapons of the future are capable of being operated and maintained by the caliber of volunteers that the services can expect to attract and retain during both good and bad economic times.

NOTES

1. The calculation is limited to males, since the number of female enlistments is constrained by legal and policy provisions. The possibility of relaxing these constraints is discussed below.
2. The number of male recruits needed each year is not solely a function of the size of the armed forces, but involves a variety of other factors such as length of enlistment, attrition rates, retention rates, and utilization of women. For simplicity, the calculation here assumed that accession needs are proportional to end strength.
3. Major General John W. Woodmansee, Jr., "Blitzkrieg and the Airland Battle," *Military Review*, August 1984, pp. 21-39.

4. See Perry's comments in U.S. Congress, Senate, Senate Armed Services Committee, *Impact of Technology on Military Manpower Requirements Hearing before Committee on Manpower and Personnel*, 96th Cong., 2d sess., 1980, p. 8.
5. Seymour J. Deitchman, *New Technology and Military Power: General Purpose Military Forces for the 1980s and Beyond* (Boulder, Colo.: Westview Press, 1979), p. 266.
6. U.S. Department of the Navy, Personnel Research and Development Center (NPRDC), *Technology Trends and Maintenance Workload Requirements for the A-7, F-4, and F-14 Aircraft* (San Diego, Calif.: NPRDC, 1979), p. viii.
7. Barry L. Scribner et al., "Are Smart Tankers Better Tankers: AFQT and Military Productivity" (West Point, N.Y.: Office of Economic and Manpower Analysis, U.S. Military Academy, 1984).
8. Conrad C. Lautenbacher, Jr., "Manning the Six-Hundred Ship Navy: Smooth Sailing or Rough Passage?" (Washington, D.C.: The Brookings Institution, 1983), pp. 4-11.
9. Tidal W. McCoy, "U.S. Armed Forces Ill-Prepared for Today's Super-Sophisticated Weaponry," *Human Events* (October 21, 1982), pp. 10-11.
10. William E. Depuy, "The All-Volunteer Force (AVF)—The Demand Side—Army Perspective," paper prepared for a conference on "The All-Volunteer Force after a Decade: Retrospect and Prospect," Annapolis, Md., U.S. Naval Academy, November 1983, p. 4.
11. The current differences between army and air force entrance standards are substantial. By best available estimates, about 77 percent of the male youth population would meet the army's minimum aptitude and education standards compared to only 63 percent for the air force. See Mark J. Eitelberg et al., *Screening for Service: Aptitude and Education Criteria for Military Entry* (Alexandria, Va.: Human Resources Research Organization, 1983), pp. 3-27, 3-30.
12. *Budget of the United States Government, FY 1986* (Washington, D.C.: Government Printing Office, 1985), pp. 3-14, 3-15.
13. U.S. Congressional Budget Office, *The Economic and Budget Outlook: Fiscal Years 1986-1990*, A Report to the Senate and House Committee on the Budget, Part I, February 1985, p. 41.
14. See statement of Robert F. Hale, U.S. Congress, Senate Armed Services Committee, *Department of Defense Appropriations for Fiscal Year 1984; Hearings before the Subcommittee on Manpower and Personnel of the Senate Armed Services Committee*, 98th Cong., 1st sess., 1984, pt. 3, p. 1650.
15. Richard L. Fernandez and James R. Hosek, "Active Enlisted Supply: Prospects and Policy Options," paper prepared for a conference on "The AVF

After a Decade: Retrospect and Prospect," Annapolis, Md., U.S. Naval Academy, November 1983, p. 2.
16. James R. Hosek, Richard L. Fernandez, and David W. Grissmer, "Active Enlisted Supply: Prospects and Policy Options" (Santa Monica, Calif.: Rand Corporation, 1984), p. 11.
17. Ibid., p. 12.
18. Gary R. Nelson, "The Supply and Quality of First-Term Enlistees under the All-Volunteer Force," paper prepared for a conference on "The AVF after a Decade: Retrospect and Prospect," Annapolis, Md., U.S. Naval Academy, November 1983, pp. 43, 49.
19. U.S. Congress, Senate Armed Services Committee, *All-Volunteer Armed Forces: Progress, Problems, and Prospects*, 93rd Cong., 1st sess., 1973, especially chapter 4.
20. Letter from Senator Warren Rudman to Colleagues, February 19, 1985.
21. Gary F. Johnson, "Rating Balance Adjustments," Alexandria, Va.: Center for Naval Analyses, July 1984.
22. For a fuller discussion of the potential for civilian substitution and of the domestic and bureaucratic politics involved, see Martin Binkin with Herschel Kanter and Rolf H. Clark, *Shaping the Defense Civilian Work Force: Economics, Politics, and National Security* (Washington, D.C.: The Brookings Institution, 1978).
23. For a description of the Hart-Toricelli proposal, see Gary Hart and Robert G. Torricelli, "Create a System of Universal National Service," *New York Times*, 14 April 1985. For discussion of the Panetta bill see U.S. Congress, Senate, *Congressional Record*, daily ed., 98th Cong., 1st sess., November 16, 1983, pp. H10026-31.
24. Martha Lynn Craver, "Senator Hollings' Bill Calls for Return to Draft," *Army Times*, 25 February 1985, p. 33.

7 MANPOWER PROCUREMENT AND MILITARY DOCTRINE
"You Can't Get There From Here"

William L. Hauser

This chapter begins with an apologia in two parts. The first is a description of the author's background, from which comes a perhaps unique insight into the problem to be discussed. The second is a listing of the author's biases, which will be discerned by the reader in the analysis anyway, so might as well be admitted beforehand. Such personalization may not be "scholarly," but it seems desirable in this case to establish credentials and to set the stage for argument.

Before serving in Vietnam, I was neither intensely aware of, nor particularly concerned about, the professional health of the armed services. I assumed, as did many of my peers, that the military had fallen into peacetime habits of bureaucratic pettiness, formalism, makework, and careerism, which it would put aside when the exigencies of war so demanded. I returned from Vietnam sadly disillusioned; if anything, those institutional ills had been magnified in the combat theater. While the army unit in which I served was perhaps a poor example,[1] subsequent discussions with contemporaries, including those in other services, convinced me that the pattern of behavior was not atypical.

Following that Vietnam experience, I spent two years in the Office of the Chief of Staff, U.S. Army. The small section in which I worked took as its highest priority the reform of the officer career system, with the premise that it was corrupting career incentives, which in turn had led to professional malaise. From that section's work came a new Officer Personnel Management System (OPMS), which sought (and still seeks) to allow officers a successful career either with troops or in technical specialization, thus avoiding the competence-diluting and behavior-distorting pressures of juggling both roles.

After a sabbatical year of research and writing on military reform,[2] I served for two years as chief of a task force studying the enlisted manpower planning and career systems. This resulted in the army's adopting a new Enlisted Personnel Management System (EPMS). In contrast to OPMS's concern with ethics, EPMS's focus was on technical competence. EPMS included six elements: (1) year-group force management, tied in with recruiting and reenlistment; (2) grouping of related military occupational specialties into career management fields; (3) rearrangement of grade structure and years-in-grade standards for promotion, to create viable career progression in each specialty and field; (4) definition of the skills required at each career level in each specialty or field; (5) in-unit or school training to provide those skills; and (6) examinations designed to measure those skills before permitting promotion. As is the case with OPMS, the army is still digesting EPMS.

After three years of duty in Germany, first as a division artillery commander and then as deputy commander/chief of staff of the army's European training command, I returned to Washington. There I held two further positions before retiring from active service. The first was a year on the Chief of Staff's Review of Education and Training for Officers. The central thrust of this study was an attempt at getting OPMS effectively implemented; unfortunately, the most important of our recommendations were either dissipated by "staffing" or sent back for "re-study." The second position was command of the Army Research Institute for the Behavioral and Social Sciences, managing investigation and analysis of recruiting and retention, training, motivation, morale and discipline, human factors engineering of materiel systems, career incentives, leadership, and organization development.

I am now director of career development for Pfizer Inc., a Fortune 100 multinational company. My principal duty is monitoring and facilitating, on behalf of the company's chairman, our operating divisions' development of executives for both the near and long term future of the corporation. It was from this position that I was asked to serve on the president's Private Sector Survey on Cost Control (the Grace Commission) in 1982-83. I headed a team on officer/NCO force management for the Department of the Army Task Force, working closely with counterpart teams on the air force, navy, and OSD task forces. That experience reinforced an earlier conclusion that the armed services are badly in need of reform in their manpower policies and career systems. I also became reluctantly convinced that while the military is incapable of fundamentally reforming itself, it is supremely capable of preventing anyone else (at least in this administration) from imposing reforms on it.

My relevant biases, to the extent I am conscious of them, can be stated in the following propositions:

1. Military doctrine is not necessarily what is written in the manuals; rather, it is what military leaders and staffs commonly understand. From that understanding, reinforced through training, doctrine becomes the way units will try to respond to situations on the next battlefield.

2. How well those units respond on that battlefield will depend on how well they have been manned, equipped, and trained before they get there.

3. The next land war is likely to begin without sufficient time to man, equip, and train units. Whatever we have in peacetime (including doctrine) is what we will have when we begin the war.

4. Whatever the intensity of that war, it will require units able to operate with tactical flexibility, with high-technology equipment, and under conditions of extreme stress.

5. In order for units to be trained to cope with such conditions at the onset of war, manpower procurement must furnish them with trainable soldiers during peacetime.[3]

6. Manpower procurement is more than simply the recruiting of entry-level people; it also includes retention of experienced professionals and hence must be viewed in the context of the entire military career system.

7. The military career system, designed for the most part in the aftermath of World War II, has outlived its usefulness and is in need of fundamental reform. Its preservation is a major impediment to the reform of other military programs and systems, including manpower procurement and tactical doctrine.[4]

THE STATE OF MANPOWER PROCUREMENT

There is a tendency, because the military career system is a closed one—that is, not permitting lateral entry—to think that manpower procurement occurs only at the bottom. In fact, there are five levels of procurement. For the enlisted force there is, first, recruitment of private soldiers; then, a few years later, from those soldiers still remaining in the service, selection of career noncommissioned officers. For officers there is recruitment at the entry level from service academies, ROTC, and OCS; then, selection into the career officer corps; and finally, selection into the managerial elite (corresponding to executives in the business world) who make policy and/or control large resources. There are no precise breaking points between each of these levels, but the processes are distinguishable. Similar processes occur in the reserve components (Reserve and National Guard) of the services.

Level 1: Recruits

Only a few years ago, the recruitment of private soldiers was in deplorable condition. In fiscal 1979, the army was able to recruit only 89 percent of its replacement needs; the navy, marines, and air force, 94 percent, 97 percent, and 98 percent respectively. Troubling as that situation was in terms of numbers, it was aggravated by the lower-than-acceptable quality of those who were recruited. In the spring of 1980, the Department of Defense revealed that of the army's 1979 (and earlier) recruits, some 45 percent were in category IV (the lowest acceptable mental level), rather than the 10 percent figure a "misnorming" of mental aptitude tests had originally led officials to believe. Additionally, new figures indicated there were correspondingly far fewer recruits in the more trainable categories I through III.[5] Although there were no figures on enlistments of category V soldiers, or on category IV and V recruits into the other services, it must be presumed that those also occurred in considerable numbers. Particularly disturbing was the revelation that pressures to fill recruitment quotas had led to widespread cases of fraudulent enlistment (e.g., ignoring of criminal records) and records falsification (e.g., phony high school diplomas).

The impact of such recruitment problems within military units was severe. There were massive failures of the army's skill qualification tests, a crucial component of the new EPMS. The other services had similar problems: navy ships that could not put to sea for lack of sailors, air force planes that could not fly sorties for lack of maintenance crews, a growing problem with discipline in the marines, as well as a fall-off in the retention and reenlistment of career officers and NCOs, in part attributable to their frustration at having to work with such low-quality troops.[6]

In the half decade since that dreadful state of affairs, the picture has become much brighter. In fiscal 1984 all of the services met their recruiting goals, both in terms of high school graduates (the best available indicator of retainability) and mental category (the best for predicting trainability in entry-level skills). The news was similarly good in terms of acquiring recruits with an aptitude for training in particular military occupational specialties. The army, habitually the service least attractive to higher quality youth, was so successful in its 1984 Delayed Entry Program (DEP), in which recruits are scheduled for entry into future training courses leading to specialties, that it planned to do all of its recruiting under DEP in 1985.[7]

But there are dark clouds on the horizon. The 1980–82 recession's aftermath of high unemployment, which so improved the quantity and quality of enlistments, has been ameliorated. Good though that may be for the country, it is bad for military recruiting, for there is a strong correlation between the unemployment of male youth and high-quality enlistments, that is, the sort that can absorb specific technical training, as in the pre-1985 DEP.[8] Indeed, some authorities predict that the American economy is on the verge of a significant shortage of technically trainable labor.[9]

There are already signs that the combination of economic expansion and baby bust is having an impact. If the army's annual requirement for high-quality (categories I through IIIA) male enlistments remains at around 65,000 as currently projected, and other conditions (pay comparability, recruiting resources, enlistment incentives, the current trend in civilian unemployment) remain constant, there will be a shortfall of more than 40 percent by 1990.[10] In a stunning reversal of its earlier optimistic stance, the Department of Defense admitted in April 1985 that youth attitude surveys revealed that the draft might soon need to be reinstated. Furious backpedaling two weeks later only confirmed the sense that the Reagan administration is genuinely worried.[11]

The issue, then, is not whether the army and the other services are going to get all the high-quality male enlistments they believe they need. They most assuredly will not. The important question is whether they really need all those high-quality young men. Could they not make up the shortfall with lower category men or with more women? Or could they not make do with more civilian employees, with physically disabled soldiers in those units requiring less vigorous activity, or with more capital-intensive materiel?

Although it is impossible to prove this in peacetime, there is probably too much reliance on civilian employees in the armed services already, given the high proportion of combat forces overseas and the likelihood that future conflict will begin suddenly. In addition to the nearly insoluble problem of what to do with overseas military families, especially the children of single parents (and dual parents both in uniform), there is much uneasiness concerning the reliability of civilian "tech reps" who now perform maintenance on high-technology equipment in overseas areas. The argument advanced in some quarters, that these people operate only in "the rear" is specious. The rearmost supply and maintenance facilities in Germany, for example, are scarcely 150 miles from the Fulda Gap, no great distance for Soviet armored blitzkrieg or parachute assault.

As for enlisting the disabled for noncombat units, the problems of defining those units (or, worse, units requiring less vigor) promises to plunge the services into unimaginable legal thickets. And the very idea that using more capital-intensive equipment would reduce the requirement for high-quality manpower flies in the face of all business experience. True, the capital intensity of equipment might move the requirement for high-quality manpower farther back in the combat theater; but, as noted above, in modern warfare of whatever intensity, there is no longer any rear.

One might think that the armed forces could use more women than they now have, if the issue were one of intelligence alone. However, much as one might dislike being labeled a "male chauvinist" for raising the issue of sexual relations, it cannot be avoided. True, the armed services have been relatively successful in integrating men and women in noncombat units during peacetime, but that proves nothing. The stresses of a combat situation will quickly reveal that sexual relationships, especially across officer/enlisted and NCO/enlisted lines, are damaging to military discipline. (Anyone who doubts that there are sexual relationships among the healthy young servicepeople in mixed-sex military units is naive as well as uninformed.) It is instructive to note that the courts have upheld the services' right to discharge homosexuals, not because homosexual relationships between consenting adults are necessarily illegal or immoral, but because sexual relationships within a military unit are likely to give rise to favoritism and emotional rivalries, and are, therefore, a hazard to discipline in a wartime setting.

There is another, even greater, oversight made by those who toss off arguments in favor of more women in military units. Units must execute doctrine on the battlefield. The unit is more than the sum of its members; it is also a network of relationships among those members. Male bonding may be an archaic, adolescent, and even reprehensible concept in the abstract, but it is a fundamental characteristic of team sports, the nearest analog to combat in our culture. Dependence on such fundamentals is common in organizations under stress. It does not take much imagination to visualize the importance of male bonding under the awful conditions of modern warfare, as exemplified by the Arab-Israeli battles of 1973 and, more recently, by the Iraqi-Iranian conflict. So, even if it were possible to recruit more high-quality women (which is difficult in a period of low unemployment), it is probably not advisable to use them in large numbers in combat-theater units. As for using more women in stateside (or shore) units, the number would be limited by the requirement to rotate troops to and from overseas and sea duty at reasonable intervals, and hence could not begin to approach the order of magnitude required by the shortfall predicted for 1990.

Finally, there is the question of recruiting and using more low-quality male soldiers. This would, at first glance, appear to be the most feasible solution. After all, cannot high-quality equipment make up for the limited ability (or, if you will, trainability) of so-called low-quality soldiers?

The answer may be found in a series of tests conducted by the Seventh Army Training Command in January through June 1984, on the tank ranges at Grafenwoehr, Germany. Two kinds of tanks were used: the M-60, workhorse of the army's tank fleet for many years, and the M-1, the newest and most high-technology armor system in the U.S. army—or perhaps in any army in the world.

The results were startling. On average, an M-60 manned by a gunner in mental category I (around the top tenth in measured intelligence) produced scores 20 percent better than one with a gunner in category IIIB (just below the median). If both the gunner and the tank commander were in category I, the scores of their tank crew were around 37 percent better than their category IIIB counterparts![12] Thus, it becomes rather obvious why the Israeli Army, relying heavily on the counsel of its psychologists, puts none but its smartest soldiers in its tank crews.

The disparity was not quite so dramatic in the case of the M-1. There, the corresponding figures were only about 9 and 10 percent.[13] Still, the results demonstrate that even the ultimately modern war machine, equipped with automatic compensations for human imperfection and operated at a firing range *where no one was shooting back*, is still significantly degraded in performance when operated by a crew of less-intelligent soldiers. It goes almost without saying that a 9 to 10 percent disparity in effectiveness (or, in business terms, productivity) on the battlefield is the equivalent of having 9 to 10 percent less manpower and weapons. Moreover, considering the stresses of the battlefield, Clausewitz's "friction of war," and the probability that many of our tanks' systems will be degraded by lack of expert maintenance, the 20 and 37 percent figures (as cited for the M-60) are the ones more likely to be experienced in reality. Thus, the case is clearly established that smarter soldiers shoot better. Recruiting more lower quality soldiers is not a viable solution to the coming shortfall.

Level 2: NCOs

As noted earlier, the second level of manpower procurement is that of providing the services with noncommissioned officers. It stands to reason that if the services have a more intelligent mix of first-enlistment soldiers, they will end up with a more intelligent mix of noncommissioned officers. The benefits to productivity are obvious; the case cited earlier for tank commanders applies equally, if not more, to ground-crew chiefs, supply petty officers, administrative technicians, engineering straw-bosses, and so on.

But there is an even more disturbing factor at work here. In times of relatively low unemployment, there appears to be an inverse correlation between the intelligence categories of reenlistment-eligible soldiers and their tendencies to reenlist.[14] So if a service wishes to have a smart NCO corps, it had better (assuming that NCOs ought to be as smart as the soldiers they supervise and lead) have an even more intelligent mix of entry-level soldiers than that called for by the soldiers' job requirements.[15] Moreover, it will have to promote these people fairly rapidly (which weakens arguments for lessening recruiting requirements by retaining an older mix of soldiers), because speed of promotion correlates positively with propensity to reenlist.[16]

In 1976, when I was about to turn over command of a division artillery to my successor, I participated in a conference organized by the senior noncommissioned officers. There were five command sergeants major present, a similar number of staff NCOs of equivalent rank, and thirty-odd first sergeants and master sergeants, a total of forty-two E-8s and E-9s.

One agenda item was a panel discussion among the five command sergeants major and myself about a project we were trying to accomplish in the command—the renewal of NCO responsibility for the performance and behavior of junior soldiers. These senior NCOs felt, as I did, that the army had slipped into bad habits of "get an officer to do it," perhaps because of a no-mistakes-allowed command climate or because of a perceived decline of overall NCO quality during the Vietnam War. Whatever the cause, the result was a self-fulfilling prophecy, a vicious circle of decreasing professional latitude and waning ability. Try as we might within our command, pressures from higher headquarters combined with the ingrained habits of our own officers and NCOs frustrated our efforts.

While there was a consensus among command sergeants major present that the army's rapid Vietnam-era expansion had degraded the quality of the NCO corps, the terms they used to describe that degeneration were not those of intelligence categories. Rather, they concerned such attributes as leadership, thirst for knowledge, judgment, attention to detail, and persistence. I asked whether these were qualities that a recruit needed to bring with him into the army, or whether they might be learned; the majority replied that the problem seemed not to be intelligence as such, but an unwillingness to learn, or an inability to concentrate on learning. (It was during the next couple of years that the army was to experience massive failures on the part of NCOs in skill qualification tests in their military occupational specialties.)[17] The division artillery operations sergeant put it this way: "It's like they don't know how to *want* to be better. I think they just weren't raised right." A grizzled first sergeant added,

"If you think these junior NCOs are bad, just wait until this current crop of volunteers puts on stripes. I'm just glad I won't be around to see it."

On impulse, I asked how many of those assembled had originally enlisted in the army and how many had been draftees. To my surprise (and continuing concern), twenty-nine of the forty-two present raised their hands to signify having been drafted into military service.

Even if it were deliberate rather than accidental, this type of "research" would not be considered valid in an experimental sense. The sample size was ridiculously small and the occasion hardly appropriate for, say, administration of a questionnaire on Armed Forces Qualification Test (AFQT) scores, parents' income, education completed, and the like. Nonetheless, the old first sergeant's prediction has come to pass: The mean intelligence score (by AFQT) of the army's NCOs in 1985, while somewhat better than in the bleak year of 1979, is significantly lower than that for 1976. More disturbing as a predictor of future senior-NCO quality is the distinct dip in AFQT scores at the E-4/5/6 level, with a mean of 48.9 compared with 54.9 and 53.8 for E-1/2/3s and E-7/8/9s respectively.[18]

Moreover, whether in recognition of quality problems or in continuation of longstanding habit, the vicious circle of NCO underutilization appears to persist.[19] This constitutes a grave tactical weakness for an army that places so much doctrinal emphasis on small-unit initiative. That weakness is probably concealed by the very underutilization that causes it, but it is nonetheless potentially disabling under what promise to be the extraordinarily challenging conditions of the future battlefield.

Level 3: Junior Officers

The third element of manpower procurement is junior officer recruitment. This is the one area that appears to be enjoying unalloyed success. The service academies have waiting lists for entrance, and ROTC is experiencing a popularity that would have been a vain dream just a few years ago. One of the causes, an upsurge of national optimism and patriotism, could be a transitory phenomenon; but the other, the high cost of a college education, appears to be here to stay. OCS programs in all the services are smaller than some would prefer, perhaps because of the soldiers who arrive with some college education,[20] but the successes of the other sources of commission make that problem less acute.

Level 4: Career Officers

When we turn to the fourth element, that of recruiting (or retaining) entry-level officers into the career officer corps, the picture is not so favorable.

The early stages of a military officer's career are extraordinarily attractive. The pay (salary plus tax-free housing and rations) is certainly comparable to what a college graduate might expect in civilian business. For example, a twenty-two-year-old O-1 makes around $19,000, a twenty-five-year-old O-2 about $26,000, and a twenty-eight-year-old O-3 perhaps $33,000. At the same time, these people enjoy an adventurous life with lots of outdoor activity, variety and novelty in their job assignments, the enrichments of travel and foreign residence, and—probably most rewarding of all—hefty responsibility at an early age. It would be almost unheard-of in the business world for employees in their mid-twenties to be entrusted with the management of a hundred or more employees and tens of millions of dollars worth of equipment, but that is what army or marine company commanders do as a matter of course.

Five years is the obligated term of commissioned service for the military academy graduate or ROTC scholarship recipient. By that time, some of them have come to realize that a military career is not what they want to do with the best years of their lives. For many of them, departure into civilian life probably cannot be prevented; after all, that is the life from which they originally came, so they are presumably making informed choices. Indeed, the services do not appear to be exerting themselves extraordinarily (except in such special cases involving pilots and nuclear-submariners) to retain junior officers.

But appearances can be deceiving. In fact, the services have been making such efforts, and on an extraordinary scale. This is a case in which the extraordinary has become invisible by being made routine. Prior to the Vietnam War, one normally spent six years in grades O-1 and O-2 before being promoted to O-3 (army/air/marine captain, navy lieutenant). The expansion of the Vietnam War reduced that figure to two years. Then, in the early 1970s, when force reductions made it likely that the period would again stretch out to its prewar span or even longer, the services made a deliberate decision to hold the line at four years. That figure was later built into the Defense Officer Personnel Management Act of 1980.

Why would the military, in a period of burgeoning social and technological change during which the end of the draft had increased dramatically the demand for mature leadership, deliberately reduce the age and experience of its army/marine company commanders and their equivalents in the air force and navy? The answer, in the absence of contrary evidence, has to be *retention.* If the services, through rapid advancement in rank and responsibility, could retain many of their better junior officers past the crucial five-year point, they might hope to keep them long enough to allow the lure of twenty-year retirement to take effect.

But do the services know how to determine who are their better officers? Performance evaluation systems are weak indicators, for they measure chiefly—at this career stage, at least—the ability to learn the military culture and abide by its rules. Not as well measured are the creative and innovative skills necessary to conceive sound doctrine and, more important, adapt doctrine to the particulars of a battle situation. The conception and adaptation of doctrine must be undertaken by a high-quality mix of officers; a few solitary geniuses cannot be effective unless reinforced by others able to respond to their ideas.

In 1970 the army's Deputy Chief of Staff for Personnel conducted the "Class of '66 Study." The members of that West Point class, essentially all of whom had seen combat in Vietnam during 1967–69, were approaching the end of their (then) four-year obligation. Disturbingly large numbers were applying to resign. The study team administered a questionnaire to those who had applied and to a control group of those who had not. In the findings the reason most often cited for resigning was the army's failure to live up to its own ideals, as these officers had been taught them at the military academy. Ignoring the significance of those data, the study concluded that those leaving were "malcontents"—and good riddance!

The point of this story is not to suggest that the army and the other services have not subsequently introduced ethical reform, for they have. Rather, it is that the services may not do very well at facing up to bad news about themselves.

I thought—hoped—to be proved wrong in that regard when I read recently of an army survey of its officers. The survey, taken over the fall-winter of 1984-85, revealed that officers, excluding generals, are disappointed in the leadership of their seniors and in the professional competence and selflessness of their peers. But when the *Armed Forces Journal* asked the Department of the Army to explain the survey's findings, the response was a discussion paper that said, in summary, "This is the best Army with the best leadership in the memory of serving officers."[21] One wonders whether to take hope in the fact that the survey was conducted, or despair that its results will apparently be ignored in yet another episode of cognitive dissonance.

Let us take another example, one having less to do with professional ideals than with the practicalities of everyday life at the officer/executive level. In the business world, there seems to be a strong correlation between the potential of younger managers and their propensity to be members of two-career marriages. The up-and-coming military officer's spouse, because of frequent reassignment (on average, every year and a half) and service overseas where career-type jobs are scarce, is severely inhibited in pursuing a career. It is altogether possible that the services are, right now, losing many of their most tactically talented young officers for this reason.

I once asked a senior officer on the Department of Defense personnel staff what was being done to address this problem. He replied that the services' school systems were designing courses to teach officers' wives how to participate more effectively in official entertainment and military community activities. When I informed him that the Grace Commission was urging reduced frequency of reassignment, for reasons of job performance and family stability as well as cost, he responded proudly that the services would succeed in ignoring such outside advice. Rapid reassignment, he said, is necessary to broaden officers for promotion; rapid promotion keeps them in the service; and twenty-year retirement both maintains promotion flow and, after a certain point, holds them longer in the service.[22]

Just how effective is twenty-year retirement in retaining high-quality officers (and NCOs)? According to the chiefs of staff of the various services, it is crucial. In testimony before the House Subcommittee on Military Personnel and Compensation on April 2, 1985, the four chiefs echoed one another. Any sort of reduction in retirement, they said, would break faith with those now serving, drive experienced officers and NCOs out of the service, so hurt recruiting as to require a return to the draft, and drastically weaken national defense. The Chief of Staff of the Army said that he had heard soldiers "voice despair" on the subject. The Chief of Naval Operations claimed that his service was "on the fragile edge."[23]

The cost of military retirement, now $18 billion a year, is predicted to rise to $45 billion (assuming there is no war) by the turn of the century. Surely these well-informed military leaders are concerned with protecting the national economy, at least to the extent of making a compromise like that recently enacted for the Social Security system. Unless one agrees with David Stockman that military leaders are "more concerned about protecting their retirement benefits than they are about protecting the security of the American people,"[24] one has to believe that military leaders are stonewalling because they do not know what else to do. As they see it, twenty-year retirement and its concomitant, "up-or-out" mandatory discharge and retirement, are crutches without which the services' manpower procurement programs would fail.

No one can predict the future, but the political handwriting on the wall is that the military retirement program is going to be drastically curtailed. It is not just that its cost is excessive (by a factor of two or three compared with programs in the business world, according to the Grace Commission),[25] but that the services can offer no truly persuasive arguments in its defense. The fact that they choose to stonewall speaks volumes about the weakness of their position. In fact, twenty-year retirement is probably *detrimental* to national security for reasons explained in the paragraphs that follow.[26]

As the end of World War II approached, the services became convinced that massive mobilization would probably need to be repeated someday. However, there would be no period of grace like the two and one-half years that elapsed from the fall of France until the American entry into offensive combat in North Africa and Guadalcanal. There would be no time for purging deadwood from the officer corps and for broadening younger officers with a variety of experiences. A career system was needed that would allow for rapid reassignment (for broad experience), frequent schooling (also for broadening), up-or-out discharge (to allow rapid promotion of better officers), and twenty- to thirty-year mandatory retirement, to keep the officer and NCO corps young and vigorous.

What the designers of the system did not foresee was that a major proportion of the armed forces would be stationed overseas, necessitating massive rotation between these assignments and assignments in the United States, and that the technological revolution begun in World War II would accelerate. Moreover, mutual nuclear deterrence with the Soviet Union would require still greater broadening of officers (at a sacrifice of depth in their skills) for instant readiness, because without such broadening even a short mobilization would be impracticable and dangerously destabilizing. Thus, efforts to keep the forces' leaders young and vigorous would greatly exacerbate assignment turbulence and inhibit officers' ability to keep up with strategic, tactical, and technological change.[27] The overall result is one repeatedly exposed by congressional and defense-intellectual members of the so-called military reform movement: a shocking decline of competence in tactical and strategic thinking and in weapons development, organizational management, and troop leadership.

The youth and vigor argument is a hollow one (except in the obvious case of combat-arms NCOs), but the services' dependence on rapid promotion, up-or-out policy, and twenty-year retirement for the survival of their manpower procurement systems is real. The fact that the services are caught in a vicious circle—albeit one of their own making—renders any partial or piecemeal solution impossible. Only a fundamental reform of the entire military career system will break the deadlock. If such reform is not undertaken, the nation will likely be faced with a continuation of recent history's dreary litany: mismanaged weapons programs, substandard readiness of the forces, repeated embarrassment over conflicts of interest, and ill-conceived military counsel to political leaders. Military fiascoes like the *Mayaguez*, "Desert One," and the Marine barracks in Beirut are harbingers of eventual large-scale military disaster.

Level 5: The Managerial Elite

As to the fifth element of manpower procurement, the recruiting of executive-level leaders and managers from among career officers, all is not well either. Most of the services admit, not officially but in their professional journals, that their officer evaluation systems are grossly inflated. As a result of this inflation (in hyperbole as well as numbers), there is a widespread belief that one black mark ruins a career. Whether this is true is not the point; if officers believe it is so, innovation, professional candor, and the willingness to take risks cannot help but be stifled. Creative solution of problems (including doctrinal problems) does not thrive in such an environment. The system may end up choosing the right people for the top, but by the time they reach the top they will have been conditioned to be risk-avoiders.

All of the problems attendant upon the five elements of manpower procurement apply also to the reserve components. Tragically, our national strategic planning, including force-structuring and tactical doctrine, is dependent on the rapid mobilization of reserves. The army provides the extreme (but not exclusive) example. Nearly 50 percent of army combat organizations are now in the reserve components; by 1989, the typical stateside division will have a reserve unit as one of its brigades. More important, over two-thirds of the army's combat support (signal, engineers, military police, etc.) and combat service support (supply, maintenance, administration) units are found in the reserves.[28]

If most of these formations were adequately trained and equipped with the same materiel as their active counterparts, then perhaps they would be deployable after minimal mobilization. But they are not so equipped and trained and, therefore, not readily deployable; indeed, some of them are manned with individuals who would automatically be deferred as defense-essential workers in time of war.[29] Non-unit members of the reserves, the so-called Individual Ready Reserve, are of equally dubious reliability, as they lack even an adequate system for enforcing submission of change-of-address cards. Repeated readiness exercises over the past decade by the Joint Chiefs of Staff have produced such consistently dismal results that the subject is too embarrassing to talk about anymore.[30]

Two conclusions may be drawn from this brief look at the manpower situation in the reserves. First, the situation is a national disgrace and would be a source of shame had it not gone on so long as to become commonplace. Second, and more germane to this paper, land warfare doctrine based upon reliance on reserve formations is a house built on sand. Any suggestion (as one sometimes hears nowadays) to meet manpower shortfalls by greater reliance on the reserves is an exercise in self-deception. It certainly will not deceive potential adversaries.

GAUGING THE IMPACT ON DOCTRINE

Let us sum up the impact of all the foregoing on military doctrine. *Doctrine* is what the units of our armed forces can be predicted to do on the next battlefield. How well they will do it depends on their state of training. That, in turn, depends on their having been manned with trainable troops, which must also be kept together long enough to learn to function as units.

Assuming that the U.S. Army (the major land force) is in fact training its units to carry out its doctrine, the question remains whether those units are being kept together long enough. It appears that the answer is no, for the army's cohesion, operational readiness, and training (COHORT) system, by its own admission, is still at only a company level of aggregation, with the companies subject to dissolution after less than two years in the overseas theater.[31] The weight of expert opinion seems to be that the U.S. military (particularly the army), despite its good intentions, lags grievously behind other modern forces in terms of building cohesion within its units.[32]

Unfortunately, we cannot even assume that the U.S. Army is training units to carry out the army's stated doctrine. First, in the current debate over the defense budget, one repeatedly hears that the allocation for operations and maintenance—in particular field training—is being sacrificed in favor of continued development of cost-escalating weapon systems. Second, there is not even agreement over doctrine itself. Our largest and most important unified command, the U.S. European Command, declines to endorse the army air force AirLand Battle doctrine and its Deep Attack corollary.[33]

So, even if our units are adequately trained at the outbreak of war, it is quite possible they will have been trained in a doctrine other than that ordered in a war setting. The result is likely to be doctrinal confusion, as units attempt to carry out intricate maneuver coordinated with multiple means of firepower, and, after failing with intricate maneuvers, a reversion to the simpler (but potentially disastrous) tactics of linear defense and attrition. To believe (as some commentators appear to) that the likelihood of these scenarios makes little difference as the possibility of war is unlikely is to ignore the maxim that deterrence derives from a credible ability to defend.

It is also necessary for battlefield units to be manned with first-rate people; for no doctrine, even if well conceived and articulated, can be executed successfully on the battlefield without a proper mix of people to put it into practice. As noted repeatedly in the army's doctrinal publications (and those put out jointly by the army and Air Force on the AirLand Battle), future tactical engagements will be characterized by intricate maneuver, enormous physical and psychological stress, high-technology weaponry, a need for small units operating autonomously and with great initiative, and a requirement for officers, noncommissioned officers, and private soldiers of high intelligence.[34] But, as discussed above, the prospects of recruiting high-quality private soldiers, retaining sufficiently high quality in the NCO corps, or developing an officer corps of excellence are not encouraging.

A PROGRAM OF REFORM

In summary, successful execution of doctrine on tomorrow's battlefield is simply not going to happen with the quality mix of officers, NCOs, soldiers, and units we are likely to have at that time. To get and keep sufficient numbers of first-rate people, the services (particularly the army) are going to have to undertake two major programs of reform.

First, there must be a restoration of the draft. The benefits will be manifold and obvious:

1. It will bring a richer mixture of intelligence (hence tactical capability) into units than will otherwise be available in the years ahead.
2. It will provide intelligent recruits to handle unit administration, thereby freeing officers and NCOs to lead their soldiers and train them tactically.
3. It will provide a richer mixture of intelligence and character for retention in the NCO corps.
4. It will make military service a respectable option for educated middle-class youth.
5. It will provide a democratic mixing of social classes in units, thereby lessening the resentment of some soldiers from disadvantaged backgrounds and from certain racial groups at having been forced by financial need to join and, in case of war, to die.
6. It will enrich and democratize the officer corps by providing numbers of college-educated enlisted people for commissioning through OCS.

7. It will provide a steady, high-quality source of manpower for the reserve components.
8. It will add to the future generation of leaders in American business, professions, and government men who have served in the military and are sensitive to the problems of national defense.

Second, as indicated earlier, there must be a fundamental reform of the officer career system. The current system, particularly its feature of early retirement, is detrimental to readiness for military operations. But even were that not the case, the system's continuation is politically unacceptable from a cost–benefit perspective. Rather than see Congress change the career system and possibly render it dysfunctional, it is in the services' interests to reform it themselves and in a manner that preserves the essence of military professionalism.

The following program would enhance the professionalism of the officer corps of the services and their readiness for military operations, as well as save billions of dollars for the country:[35]

1. Continuation of an up-or-out policy for officers and NCOs short of 20 years' service.
2. Continuation of 20-year, immediate-pension, voluntary retirement for combat-arms NCOs.
3. Replacement of immediate pensions for officers and noncombat-arms NCOs with 20 to 25 years' service with a combination of lump-sum severance pay and a deferred annuity. For those with 25 to 30 years of service, a proportionally larger severance and annuity package plus an intensive career-transition program lasting for several months while they remain on active duty and at full pay and allowances. For those with 30 or more years of service, a choice of immediate pension or the combination of severance, annuity, and outplacement.
4. Agreement to a minimum-tenure assignment "contract" for certain officers with 20 or more years' service—those chosen for important positions requiring continuity for effectiveness. Such agreement will be required before appointment, and subsequent adherence to it will give these officers extra severance pay and outplacement (even for 20 to 25 years), while non-adherence will reduce their severance and deny them outplacement.
5. Retention of selected officers and NCOs to 35 years in areas where excellence is required and stability and maturity are in-

valuable: recruiting, ROTC teaching, Reserve and National Guard advising, administration of troop training centers, and—for a few carefully selected colonels/captains and generals/admirals—detachment from parent service and duty on the Joint Staff up to 40 years' total service.

These reforms, or some variation of them, need to be made soon. Their impact on the quality of career officers and NCOs and, therefore, the performance of units and staffs on future battlefields will require years to take full effect. Whether or not one agrees with these particular reforms, it is indisputable that "we can't get there from here" on the road we are now traveling. Some drastic changes are needed. Let us hope we can summon up the will to make them, and that the march of events gives us time to get the job done.

NOTES

1. The behavioral and operational absurdities found in the U.S. Army's 9th Infantry Division in 1968-69 have been described brilliantly by then-Major Josiah Bunting in *The Lionheads* (New York: Braziller, 1972), and unwittingly by Lieutenant General Julian J. Ewell and Major General Ira A. Hunt, Jr., in their official report, *Sharpening the Combat Edge: The Use of Analysis to Reinforce Military Judgment*, Vietnam Studies Series, Department of the Army (Washington, D.C.: 1974). Edward N. Luttwak, in *The Pentagon and the Art of War: The Question of Military Reform* (New York: Simon and Schuster, 1985), p. 31n., calls the latter "an exercise in unconscious self-parody."
2. William L. Hauser, *America's Army in Crisis: A Study in Civil-Military Relations* (Baltimore: Johns Hopkins University Press, 1973).
3. The doctrine discussed in this paper is that pertaining to land warfare, hence the frequent use of the terms *soldiers* and *units*. The author's scant knowledge concerning both sea and air warfare, except for those aspects directly supporting land forces, is readily conceded.
4. That may well be thought a predictable approach by a professional career developer, for "To the man with a hammer, every problem looks like a nail." However, these views were essentially formulated fifteen years ago ("Professionalism and the Junior Officer Drain," *Army*, September 1970, pp. 16-22), before the author began pursuing the career-development business.
5. Franklin D. Margiotta, "Evolving Strategic Realities," in Franklin D. Margiotta, ed., *Evolving Strategic Realities: Implications for U.S. Policy-*

    makers (Washington, D.C.: National Defense University Press, 1980), p. 102.
6. Ibid., p. 104.
7. General Maxwell Thurman, Vice Chief of Staff, U.S. Army, address to the Association for a Better New York, New York City, January 16, 1985.
8. Charles Dale and Curtis Gilroy, *The Economic Determinants of Military Enlistment Rates*, Technical Report 587 (Alexandria, Va.: U.S. Army Research Institute for the Behavioral and Social Sciences, 1983), p. 2.
9. For example, John Diebold, *Making the Future Work: Unleashing Our Powers of Innovation for the Decades Ahead* (New York: Simon and Schuster, 1985), p. 20.
10. David K. Horne, *An Economic Analysis of Army Enlistment Supply*, Working Paper MPPRG 84-5 (Alexandria, Va.: U.S. Army Research Institute for the Behavioral and Social Sciences, 1984), pp. vii, 39.
11. Richard Halloran, "Enlistment Decline Brings Call for New Draft," *New York Times*, 9 April 1985, p. A1; and "Draft Won't Be Needed to Fill Ranks, Senior Military Officials Say," *New York Times*, 23 April 1985, p. A23.
12. Barry L. Scribner et al., "Are Smart Tankers Better Tankers: AFQT and Military Productivity" (West Point, N.Y.: Office of Economic and Manpower Analysis, U.S. Military Academy, 1984). See also Lieutenant Colonel Thomas W. Fagan, "An Agenda for the Secretary of Defense— 1985 Personnel Costs" (West Point, N.Y.: Office of Economic and Manpower Analysis, U.S. Military Academy, 1985). Similar conclusions (that high-quality operators are essential to optimum performance of even the most modern weapon systems) may likewise be drawn in the case of officer crew members, as reflected in the army's decision to place only one pilot in the new LHX helicopter. See Major General Carl H. McNair, Jr., "LHX: The Conceptualization of Why, What and When," *Army Aviation Digest*, December 1982, pp. 2-5; and Lieutenant Colonel W. Lawson, "Light Helicopter Family (LHX): The Army's New Family of Light Rotorcraft," Department of the Army briefing, 1984.
13. Scribner et al., "Are Smart Tankers Better Tankers."
14. Hyder Lakhani and Curtis Gilroy, *Army Reenlistment and Extension Decisions by Occupation*, Working Paper MPPRG 84-14 (Alexandria, Va.: U.S. Army Research Institute for the Behavioral and Social Sciences, 1984), pp. 34-35.
15. *Report to Congress on Army Enlisted Manpower Quality Requirements* (Washington, D.C.: Department of the Army, 1985), Executive Summary, p. iii. See also the similarly titled reports of the navy, air force, and marines. See also *Manpower Requirements Report FY 1985*, Vol. 3 (Washington, D.C.: Department of Defense, 1984), and *Manpower Requirements Report FY 1986*, Volume 3 (Washington, D.C.: Department of Defense,

1985). It is interesting to note that the 1986 volume omits the 1985 volume's section on "Manpower Readiness."
16. Robert H. Baldwin and Thomas V. Daula, "Speed of Promotion, Military Pay and the Reenlistment Decision: An Empirical Investigation" (West Point, N.Y.: Office of Economic and Manpower Analysis, U.S. Military Academy, 1984), pp. 5, 27.
17. Margiotta, *Evolving Strategic Realities*, p. 104.
18. Raw data obtained from the U.S. Army Recruiting Command, Fort Sheridan, Ill., May 1985.
19. *Soldiers Report III, 1984* (Washington, D.C.: Department of the Army, 1984), pp. 3-5–3-10.
20. Charles C. Moskos, Jr., "The Enlisted Man in the All-Volunteer Army," in *Manpower Resources Management: Readings for Academic Year 1984–85* (Washington, D.C.: Industrial College of the Armed Forces, 1984), sec. E-3, pp. 2-3.
21. Benjamin F. Schemmer, "Internal Army Surveys Suggest Serious Concerns About Army's Senior Leaders," *Armed Forces Journal*, May 1985, pp. 18–20.
22. Deputy Assistant Secretary of Defense for Manpower and Reserve Affairs, interview with William L. Hauser, 13 September, 1982, Washington, D.C.
23. Richard Halloran, "4 Military Chiefs Oppose Any Change in Pensions," *New York Times*, 3 April 1985, p. A18. Since that article was written, the Pentagon has softened its position, in quick response to a vote by the House Armed Services Subcommittee on Military Manpower and Compensation curtailing funds for retirement. Halloran, "Pentagon Shifts on Pension Trims," *New York Times*, 5 May 1985, pp. A1 and A27.
24. Halloran, "4 Military Chiefs."
25. President's Private Sector Survey on Cost Control, *Report on the Office of the Secretary of Defense* (Washington, D.C.: Government Printing Office, 1983), p. 213.
26. For a more detailed analysis, see William L. Hauser, *Restoring Military Professionalism*, Heritage Backgrounder, No. 449 (Washington, D.C.: Heritage Foundation, 1985).
27. To be fair, that is what the army's OPMS is trying to correct by allowing officers to narrow their career patterns without sacrificing success. In like manner, the navy just this year announced a program to give supply and materiel specialists more viable career opportunity. "Navy Opens Slots for Materiel Pros, Disbands Naval Materiel Command," *Armed Forces Journal*, May 1985, pp. 20, 22.
28. Colonel Charles R. Hansell, "Army Reserve Component Status and Potential," in *Reserve Component Manpower Readiness and Mobilization Policy*, ed. Barbara A. Henseler, Hardy L. Merritt, and James L. Gould (Washington, D.C.: Mobilization Concepts Development Center, National De-

fense University, 1984), vol. 2, pp. K-1-1–K-1-8. See also Lieutenant Colonel Robert E. Keenan, Jr., "Army Modernization and Mobilization: Friends or Foes?" in *Reserve Component Manpower Readiness and Mobilization Policy*, eds., Henseler, Merritt, and Gould, vol. 2, pp. P-6-1–P-6-12. See also *Fiscal Year 1983 Readiness Assessment of the Reserve Components* (Washington, D.C.: Reserve Forces Policy Board, Office of the Secretary of Defense, 1984), pp. EX-1–EX-9.

29. Kenneth J. Coffey, *Manpower for Military Mobilization* (Washington, D.C.: American Enterprise Institute, 1978), pp. 41–47.

30. Captain William L. Bolton, "Mobilization of Guard and Reserve Forces and the Legal System," in *Reserve Component Manpower Readiness and Mobilization Policy*, eds., Henseler, Merritt, and Gould, pp. D-1-1–D-1-9.

31. Lieutenant General Robert M. Elton, "Cohesion and Pride Aims of New Manning System," *Army*, October 1984, pp. 218–228. It is revealing to note that, in the same issue, neither the Deputy Chief of Staff for Operations and Plans nor the Commanding General of the Training and Doctrine Command thought the new manning system significant enough, tactically or doctrinally, to mention in their respective articles.

32. Defense Management Study Group on Military Cohesion, *Cohesion in the U.S. Military: An Industrial College of the Armed Forces Study in Mobilization and Defense Management* (Washington, D.C.: Industrial College of the Armed Forces, 1984), pp. ix–xvi. See also Colonel William Darryl Henderson, *Cohesion: The Human Element in Combat* (Washington, D.C.: National Defense University Press, 1985), pp. 151–60. See also Lieutenant Colonel John F. Guilmartin and Lieutenant Colonel Daniel W. Jacobowitz, "Technology, Primary Group Cohesion, and Tactics as Determinants of Success in Weapons System Design: A Historical Analysis of an Interactive Process" (Montgomery, Ala.: USAF Air Command and Staff College, 1984).

33. Senior NATO military officer interview with author, 16 April 1985, Washington, D.C.

34. Department of the Army, *Report to Congress*. See also Army Field Manual 100-5, *Operations* (1982), Army Field Circular 100-1, *The Army of Excellence* (1984), and Colonel John R. Landry et al., "Deep Attack in Defense of Central Europe: Implications for Strategy and Doctrine," in *Essays on Strategy* (Washington, D.C.: National Defense University Press, 1984), pp. 29–78.

35. A detailed explanation of this program may be found in Hauser, "Restoring Military Professionalism."

V PROSPECTS FOR A FUTURE SYNTHESIS

8 IMPLICATIONS OF LIKELY FUTURE CONFLICT ENVIRONMENTS FOR U.S. MILITARY MANPOWER POLICIES AND PRACTICES

Jeffrey Record

This chapter postulates future conflict environments that are most likely to engage U.S. military power and then assesses the broader implications of those environments for U.S. military manpower policies and practices. The first of these tasks is inherently the more difficult, since it necessarily entails a large measure of speculation. Certain informed conclusions can nevertheless be drawn, given the history of conflict since 1945, the character and locus of present and likely future threats to U.S. security interests overseas (the United States has conducted no major conventional military operations on the North American continent since the Mexican–American War of 1848), and the persistent deterrent power of nuclear weapons.

It is assumed, for example, that the superpowers will continue for the foreseeable future to refrain from using nuclear weapons against one another, and that major conflict in Europe and in Northeast Asia involving the United States will continue to be deterred by the presence of nuclear weapons. It is further assumed that both interstate and intrastate violence will continue to characterize the resolution of political disputes in much of the so-called Third World, particularly Africa, the Middle East, and Southeast Asia, as has been the case for the past forty years.

THE SPECTRUM OF CONFLICT

Today, and for the foreseeable future, the spectrum of potential conflict environments in which the United States could find itself engaged ranges in intensity (scale of violence) from general nuclear war to what for lack of a better term is here labeled *incidents*, or individual acts of violence directed against the military forces, nationals, or property of one or more states. Past examples of the latter include North Korea's seizure of the USS *Pueblo* in 1968, the seizure of American diplomatic personnel in Tehran in 1979, and the terrorist bombing of U.S. Marine Corps headquarters in Beirut in 1983.

Between these two extremes lie several categories of conflict, which in descending order of intensity are: limited nuclear war; general, or worldwide, conventional war; limited conventional war (e.g., the Korean, Vietnam, and Falklands wars); and small, semiconventional and unconventional wars and military actions, usually characterized by small-unit operations employing guerrilla, terrorist, or other irregular tactics and directed against foreign or indigenous state forces. This last category includes intrastate rebellions, revolutions, and so-called wars of national liberation, with or without the support of outside powers (e.g., the Chinese Civil War, the Algerian War, the Iranian Revolution, and the present war in Afghanistan).

It is to be noted that most wars since 1945, all of which have been non-nuclear, have contained elements of both conventional and unconventional combat.

CONFLICT ENVIRONMENTS SINCE 1945

The most distinguishing feature of war since 1945 has been its limited intensity and geographic scope. There has, moreover, been a distinct correlation between intensity and occurrence; whereas comparatively low-intensity, unconventional conflict has been a constant feature of the international political environment, mid- to high-intensity conventional combat has occurred less frequently. Notable for its complete absence has been nuclear conflict and general conventional war.

Indeed, for the first time in modern history the great powers, which for our purposes are defined as those states possessing the capacity to retaliate in kind (if not proportionally) to a nuclear attack, have altogether refrained from making *any* kind of war *directly* against one another, preferring instead to engage one another indirectly through client states and surrogate forces. And since the great powers are the only states capable of waging either nuclear war or general conventional war, their refusal to do so has restricted the scope of war in the nuclear age.

The deterrent power of nuclear weapons has not only confined conflict since 1945 to and below the level of limited conventional war, but also eliminated Europe as an arena of combat. For centuries a hotbed of interstate violence and the cockpit of both world wars, Europe has remained an island of comparative peace (the major exception being the Soviet Union's occasional invasions of its East European allies). And it is difficult to envisage a set of circumstances in the future in which either NATO or the Warsaw Pact would see itself compelled to initiate major hostilities on the Continent. For both superpowers, a war in Europe would risk catastrophic vertical and horizontal escalation far outweighing any conceivable political objectives for which war might be initiated.

None of this is to argue that war in Europe is an impossibility. Neither World War I nor World War II succeeded in resolving the German question; Europe remains an armed camp, and the possibility of war through accident or miscalculation can never be completely dismissed. Nor is it to argue that preparation for war in Europe can or should be abandoned as the principal planning determinant of the size and character of U.S. general purpose forces; the absence of war in Europe for the past forty years is attributable in no small measure to the heavy U.S. investment in Europe's defense. It is simply to recognize that the possibility of war in Europe is remote, certainly as compared to other conflict environments that have engaged U.S. military power and are likely to do so in the future.

Indeed, the deterrent power of nuclear weapons has conditioned the engagement of U.S. military power no less than it has the military power of other nuclear-armed states. Since 1945 the United States has fought two sizable wars (in Korea and Indochina), both of them outside Europe and against non-nuclear adversaries. Additionally, in both cases the United States, while reportedly contemplating the use of nuclear weapons, not only refrained from doing so but also undertook positive measures aimed at containing the level and geographic scope of conventional violence. All other significant applications of U.S. military power that involved or resulted in some measure of violence—the Dominican Republic intervention of 1965, the *Mayaguez* rescue mission of 1975, the Iranian hostage rescue mission of 1980, the deployment of Marine Corps forces to Lebanon in 1982, and the invasion of Grenada in 1983—also were undertaken outside Europe against non-nuclear adversaries.[1]

These recent applications of U.S. military power were, moreover, undertaken for one or more of the following three reasons: (1) to halt aggression by a local communist state against an ally or client state of the United States; (2) to prevent the overthrow of allied or friendly governments by indigenous or outside forces perceived to be hostile to the United States and/or its regional security interests; or (3) to recover American nationals kidnapped or otherwise placed in jeopardy by hostile states or state-sponsored terrorist organizations. These requirements are likely to continue to predominate in the future.

In terms of operational environments encountered, all violent applications of U.S. military power shared a number of features. All were non-nuclear, most of them of mid- to low intensity. All were conducted at considerable distance from the United States, placing a premium on strategic mobility and on forcible-entry capabilities and/or secure military access ashore in the theater of operations. Even the comparatively proximate Dominican and Grenadan actions required projection of military power over hundreds of miles of water. Moreover, most applications were conducted in rugged, heavily foliated, relatively roadless, and/or otherwise close terrain that severely constrained target acquisition and surface mobility and precluded decisive applications of airpower.

No less significant was the character of enemies encountered. All were primarily foot-mobile infantry formations or terrorist groups, albeit often highly disciplined, trained, and well armed; operating for the most part in an unconventional manner;[2] and lacking attendant naval and air power capable either of disrupting U.S. ground operations or U.S. strategic lines of communications. Not since 1945 has the United States done battle with a heavily mechanized adversary in conditions suitable for large-scale armored warfare, although armored warfare continues to thrive in the Middle East, and the balance of the U.S. Army's sixteen divisions are fully mechanized. Nor has U.S. tactical air power ever been successfully challenged in the post-World War II era.[3] Even more notable has been the absence of classical naval engagements. To be sure, U.S. naval power was employed extensively in both the Korean and Vietnam wars, but it was employed exclusively in support of ground and air operations. Indeed, the only significant World War II–style naval engagements to occur anywhere since 1945 were those that took place off the Falkland Islands in 1982.

In sum, although U.S. military forces since 1945 have remained postured primarily for high-intensity conflict against Soviet forces in Europe, the actual engagement of U.S. military power has been confined exclusively to mid- to low-intensity conflicts against non–Soviet adversaries outside Europe.

FUTURE CONFLICT ENVIRONMENTS

That mid- to low-intensity conventional warfare has been the predominant conflict environment of the past forty years does not necessarily mean that it will predominate in the future; the future is never simply an extension of the past. On the other hand, there are reasons for believing that future conflict environments, particularly those most likely to engage U.S. military power, will remain nonnuclear and confined to the level of limited conventional war and below. Indeed, a strong argument can be made that lesser intensity semiconventional or unconventional war is likely to emerge as the predominant conflict environment for U.S. military forces. Such has been the case since the last U.S. ground combat forces were withdrawn from Vietnam in the early 1970s.

Barring a highly unlikely transformation of the U.S.-Soviet nuclear balance that would afford one side or the other a decisive superiority,[4] nuclear weapons will likely continue to deter direct conflict among the great powers as well as any major conflict in Europe. This means that, as in the past, the United States and the Soviet Union will continue to seek and preserve strategic advantage over one another indirectly, through client states and surrogate forces, and that this contest, to the extent that it erupts in violence, will be conducted largely in Third World areas peripheral to centers of direct U.S.-Soviet military confrontation.

But Soviet and Soviet-sponsored violence in strategically vital areas of the Third World is neither the only nor perhaps even the greatest long-term threat to American security interests overseas. During the past two decades or so there has arisen in the Third World a new category of remarkably virulent threats to those interests, the distinguishing features of which are their fanatical motivation, indigenous origins, and unconventional style of violence. Largely the product of rabidly anti-Western (and especially anti-American) religious and cultural environments native to certain areas of the Third World (e.g., Khomeini's Iran, Qadaffi's Libya, and Pol Pot's Kampuchea), these threats have manifested themselves at the very lowest ends of the conflict spectrum, mainly in terrorist attacks against U.S. military forces and individual citizens overseas.

Although space does not permit a detailed analysis of mounting international terrorism and the threat it poses to U.S. security interests overseas, suffice it to say that U.S. general purpose military forces are not organized, equipped, or trained to deal effectively with such threats. As the ill-fated U.S. military intervention in Lebanon in 1982–1983 demonstrated, conventional military responses to terrorism are more or less irrelevant. And as the debacle of the Iranian hostage rescue mission revealed, even special operations, if misconstrued as simply conventional operations writ small, are also likely to fail.

Aside from terrorism, the future conflict environment most likely to engage U.S. military power is that of small or medium-sized semiconventional or subconventional wars stemming from externally supported indigenous attempts to overthrow politically unstable Third World governments allied or friendly to the United States. The major current example of such a conflict is the guerrilla war in El Salvador, sponsored by the neighboring Sandinista regime in Nicaragua. Indeed, recent events in Central America and the Philippines, as well as the continued deterioration of sub-Saharan Africa's political and economic environments,[5] suggest a distinct requirement for both small-scale, Grenada-style actions and larger, more prolonged military operations in these and other areas of the Third World, including the Arabian peninsula, where threats to local U.S. friends and clients are fundamentally internal rather than external. As of the end of 1984, the most likely arena of such operations appeared to be Central America and the Caribbean. Ongoing U.S. concern over continued Soviet-Cuban attempts to destabilize both areas not only provoked a U.S. invasion of Grenada in 1983 but also raised the very real possibility of future direct U.S. military intervention in Central America and even against Cuba itself.

Next in order of likely future conflict environments is a geographically limited, high- to mid-intensity conventional conflict against large, well-equipped, Soviet-model client militaries. Recent events along the Thai-Cambodian border and in the upper Persian Gulf have underscored the continuing vulnerability of important U.S. client states and of future U.S. and Western access to the indispensable energy resources of the Persian Gulf and associated lines of communication. Direct U.S. engagement in such conflicts in either area, however, must be regarded as relatively improbable. There is no serious evidence that North Vietnam seeks to extend the borders of its own small empire beyond those of old French Indochina, and any attempt by Hanoi to do so probably would elicit an armed Chinese response far larger than that which followed North Vietnam's invasion of central and western Cambodia in the late 1970s. In the case of the Persian Gulf, threats to Saudi Arabia and other conservative states on the Arabian peninsula are, as noted, mainly internal and unconventional in nature; and Syria, the region's most militarily powerful potential adversary of the United States, shares no border with any state on the peninsula and is likely to remain militarily preoccupied by events in Lebanon for the foreseeable future.

The likelihood of a Soviet invasion of Iran, which remains the "worst case" force planning focus of the U.S. Central Command (CENTCOM) is even more remote. In any event, both the feasibility and the desirability of a U.S. defense of Iran against a Soviet attack, should one materialize, are so questionable as to cast suspicion on U.S. CENTCOM's motives.[6] A successful U.S. defense of Iran against a determined Soviet invasion would require the following conditions, few if any of which are realistically obtainable: (1) substantial warning and an early decision to act upon that warning; (2) unfettered U.S. military access to air facilities in Turkey and Saudi Arabia for the purpose of staging operations against Soviet forces in Iran; (3) Iranian acquiescence to the presence of U.S. combat forces on Iranian territory; (4) an ability to keep the Strait of Hormuz open against Soviet subsurface, surface, and air attacks, including *Backfire* bomber attacks delivered directly from bases inside the Soviet Union; and (5) a refusal on the part of the Soviet Union, despite its disposition of larger forces against both Iran and Western Europe and its possession of interior lines of communication between the two theaters, to engage in any horizontal escalation of the conflict.

Given the current and probable future character of the Iranian political leadership, U.S. military action in Iran is far more likely to be directed against Iranian military forces and installations than against Soviet forces.

ASPECTS OF FUTURE CONFLICT ENVIRONMENTS

The foregoing discussion suggests that for the foreseeable future U.S. military power is likely to be predominantly engaged, as was British military power in the latter half of the nineteenth century, in distant small-scale wars and punitive military actions of varying duration against local non-Western adversaries.[7] Unlike the British, however, the United States is not likely to enjoy a crushing technological or doctrinal advantage over its opponents; Third World access to modern armaments and to sophisticated Western and communist organizational concepts and tactical doctrines has eliminated the kind of decisive superiority once conferred by the Maxim gun over the *assegai*.

Nor did "Mr. Kipling's Army" have to contend with acute public sensitivity to long wars and long casualty lists. The days are long gone when any Western democracy can take 60,000 casualties on a single day, as did the British on the Somme in 1916, and come back for more with bands playing and flags waving. Indeed, sensitivity to casualties has come to exert a prominent influence on Western force planning and military operations. The United States was driven from Vietnam in part by public reaction to casualty rates, notable historically only for their relative insignificance (American battle deaths per month in Vietnam averaged 678, compared to 8,929 during World War II; British and French monthly battle deaths during World War I averaged, respectively, 17,734 and 26,315). The strategy, tactics, and tempo of Britain's recapture of the Falkland Islands in 1982 and of the subsequent U.S. invasion of Grenada were shaped by acute governmental appreciation of public intolerance of long casualty lists, especially in distant conflicts in which vital security interests are not perceived to be at stake. And it was public and congressional shock over the deaths of 241 Marines at the hands of a lone terrorist in Beirut that ultimately compelled a withdrawal of U.S. military forces from Lebanon.

In terms of geographical loci, Central America, Southwest Asia, and Southeast Asia would seem to be the most likely arenas of future U.S. military action. All three areas are characterized by predominantly rugged, mountainous, heavily foliated, comparatively roadless, or otherwise close terrain severely inhibiting, and in many cases precluding, a decisive application of mechanized ground forces and tactical air power.[8] This is especially true of Central America and Southeast Asia, where both weather and terrain place a premium on hearty foot-mobile infantry and fire support (including helicopter gunships) organic to engaged ground forces.

To these natural constraints on ground-force tactical mobility and firepower delivered by fixed-wing aircraft must be added those likely to be deliberately imposed by potential adversaries of the United States in Central America, Southwest Asia, and Southeast Asia. As the Vietnam War, the Lebanese intervention, and the current conflict in Central America all demonstrate, terrorist groups, as well as military forces operating in an unconventional manner (i.e., in a manner designed to offer minimal exposure to conventional applications of firepower), can more than hold their own against—and often beat—larger, heavier, and better equipped regular forces. This is, in fact, the tactical essence of terrorist and guerrilla operations, and the past success of such operations against U.S. and client military forces in all three theaters of operations under discussion suggests that this style of warfare, conducted mainly by small units taking advantage of natural cover (and of inadequate U.S. intelligence, especially concerning terrorist organizations operating in the Middle East), is likely to remain the preferred style of prospective U.S. adversaries in all three theaters.

In the case of Central America, among the more likely arenas of future U.S. military operations, no potential U.S. adversary even possesses the ability to mount significant cross-border conventional military operations. Notwithstanding its continuing acquisition of tanks, armored fighting vehicles, and other conventional arms and equipment from the Soviet bloc,[9] Nicaragua's military is thoroughly guerrilla in outlook and lacks any experience in conventional military operations. More to the point, such operations would be severely inhibited in the restricted terrain of Central America (Nicaragua shares no border with El Salvador, and the only all-weather road connecting the two countries runs through Costa Rica).

None of this is to discount the possibility of large-scale conventional military operations. That possibility is inherent in the large, highly mechanized Soviet-model armies of Syria, Iraq, and Korea, among others.[10] The U.S. military, by virtue of its continuing force-planning focus on deterring a Soviet invasion of Europe, would seem comparatively well prepared to deal with that possibility (assuming there is sufficient warning and adequate strategic mobility). Informed judgment, however, argues strongly for the greater likelihood of smaller scale semiconventional and unconventional conflict characterized by those features discussed above.

IMPLICATIONS FOR U.S. MILITARY MANPOWER POLICIES AND PRACTICES

If valid, the postulation of small, semiconventional and unconventional wars as the most likely future conflict environments to engage U.S. military forces contains profound implications for Pentagon manpower policies and practices, especially those of U.S. ground forces. As noted, U.S. ground and tactical air forces remain postured primarily for high-intensity combat against Soviet forces in the European theater, and this is unlikely to change, barring a fundamental redistribution of the military burden of Europe's defense within NATO. Unlike Great Britain in the latter half of the nineteenth century, the United States has profound security interests in Europe that can be preserved only by a major force presence on the Continent; it does not have the luxury of tailoring its forces exclusively for only one or two kinds of war.

It is worth recalling, however, that the small-war army fielded by Great Britain during that period was a small, elite, volunteer force of long-service professionals, many of them recruited locally throughout the British Empire, characterized by superb training and unexcelled levels of small unit cohesion, the latter the product of a regimental system that more or less kept the same officers and men together throughout their military service. It was also an army composed primarily of light infantry, light cavalry, and light artillery units, whose agility and logistical austerity allowed them to operate effectively in the distant, often remote, and infinitely varied operational environments of the empire. Equally notable was a structure of command, distinguished by decentralization and encouragement of junior officer/small-unit initiative, that evolved not only because of the difficulties distance and primitive communications engendered but also as a result of a recognition of the unique requirements of small wars waged in unfamiliar and distant environments.

Promotion of these values—smallness, lightness, specialization, quality of training, unit cohesion, and decentralization—would seem no less imperative for the small-war conflict environments of the future than they were for the small-war environments of the past. Small wars in obstructed terrain against elusive enemies place a premium on small, light, specialized, cohesive, elite, and tactically flexible forces, as well as on forces (given Western political culture's increasing sensitivity to casualties and long wars) capable of getting the job done quickly and at minimal cost. The Grenada operation of 1983, which was conducted by elite light forces and concluded in a few days with little loss, would seem a model for the future, although the unusually favorable circumstances bearing on that operation are unlikely to recur in the future. In terms of U.S. manpower policies and practices, likely future conflict environments would appear to place a premium on the following values: quality rather than quantity, specialized training and the revival of warrior values, small-unit cohesion, and decentralized command authority.

Quality Rather Than Quantity

Small wars by definition require smaller forces than large wars. More to the point, small wars waged in rugged natural environments and against evasive opponents require a level of physical stamina, independent judgment, and initiative greater than that traditionally associated with mass combat between mechanically mobile forces. As the Falklands War demonstrated, ground forces operating in distant, rugged, and climatically extreme theaters of conflict, and under circumstances in which most modes of tactical mobility other than marching are denied, must maintain exceptional standards of physical fitness as well as enjoy a wide latitude in tactical and operational decisions.

Whether the all-volunteer force can continue to recruit and retain sufficient numbers of physically robust and mentally qualified people is far from certain. There are already signs that the recruiting and retention boom of the past half decade may be coming to an end.

Continuing economic recovery has reduced the Pentagon's competitiveness in the civilian manpower market, and inexorable adverse demographic trends have led many analysts to conclude that a return to conscription is inevitable before the end of the century. Certainly, as long as manpower requirements for small wars outside the European theater must compete with those derived from preparation for major conflict in Europe, the issue of manpower quality with respect to likely future conflict environments will remain problematic.

Specialized Training and the Revival of Warrior Values

Likely future conflict environments also place a premium on specialized rather than general purpose training, and on traditional warrior values rather than managerial-technocratic values. The natural environments of Central America, Southwest Asia, and Southeast Asia differ widely from those of Central Europe. Mountains, deserts, and jungles impose unique demands on military operations, and warfare in each requires rigorous, lengthy, specialized training as well as specialized force structures and equipment. Mastering the art of these and other unique forms of warfare, including cold-weather and amphibious warfare, requires far more time than that needed to gain minimal proficiency in general purpose warfare. This is particularly true of special operations, the most demanding of all forms of small-unit military operations—certainly the most indispensable capabilities for dealing with unconventional threats of the terrorist variety, yet, regrettably, the form of warfare in which the United States has been least successful.

U.S. ground forces today contain few units—and none of division size—specialized for mountain, desert, or jungle warfare, despite the fact that one or more of these three natural environments predominates in Central America, Southwest Asia, and Southeast Asia. The U.S. Marine Corps predictably remains wedded to amphibious warfare, an essential element of U.S. force projection capabilities, although of limited value for sustained inland combat. And the U.S. Army has, at least until recently, exhibited a preference for large, general-purpose force units optimized for prolonged non-nuclear combat conducted in a conventional manner.

The demands exerted by the small wars likely to engage future U.S. military forces will also place a premium on the restoration of traditional warrior values within the officer corps, values that in the past two decades have been largely displaced by the managerial-technocratic values endemic to the vast Pentagon bureaucracy.

Unlike many large wars, such as the American Civil War and World Wars I and II, small wars are not battles of materiel, decided by weight of resources. Small wars are manpower, not capital, intensive. As such, like the Vietnam and Falkland Wars, their outcome is not determined by either technological superiority or a superiority in the measurable indices of military power but rather by such time-tested intangibles as operational and tactical skill, leadership, esprit, training, cohesion and the like—in short, the traditional values of the warrior.

It is not within the purview of this chapter to prescribe specific measures for the restoration of these values within the U.S. military. Suffice it to say that their restoration will entail significant alterations in, among other things, the current military educational system and curricula, which today emphasize administrative technique and technical expertise at the expense of a thorough knowledge of military history; in service promotion policies, which today all too often reward careerism rather than professional competence; and in service personnel management systems, which today corrode unit cohesion by continually shuffling officers and men from unit to unit and job to job.

Small-Unit Cohesion

The Vietnam War demonstrated the folly of conventional, large-unit operations against an unconventional adversary in close terrain, and both the Vietnam War and the Iranian hostage rescue mission demonstrated the penalties for failing to maintain proper levels of small-unit cohesion in the face of the stress, chaos, and uncertainties of combat.

Small wars are essentially collections of often prolonged and intense small-unit actions under unusually rigorous physical and climatic conditions, all of which place an unusual premium on the moral and social cohesion of individual battalions and companies.

And there is only one way that that cohesion can be created and maintained. As Ardant du Picq, himself a veteran of France's small wars in Africa and the Middle East during the middle of the nineteenth century, wrote shortly before his death in 1870:

> A wise organization insures that the personnel of combat groups changes as little as possible, so that comrades in peacetime maneuvers shall be comrades in war. From living together, and obeying the same chiefs, from commanding the same men, from sharing fatigue and rest, from cooperation among men who quickly understand each other in the execution of warlike movements, may be bred brotherhood, professional knowledge, sentiment, above all unity.[11]

Unfortunately, as noted, current service-manpower management systems do not ensure "that the personnel of combat groups changes as little as possible"; quite the opposite is the case. Further endangering small-unit cohesion is the U.S. military's continued reliance on the individual replacement system, whose inferiority to the unit replacement system as a means of sustaining cohesion on the battlefield was glaringly evident as far back as World War II.[12] For U.S. ground forces, however, even the adoption of a British-style regimental system and unit replacement might be insufficient to provide desired levels of small-unit cohesion. Minimum terms of enlistment in the present all-volunteer force range from two to four years (in contrast to seven years for the volunteer British Army), which may not be adequate to breed the necessary "brotherhood."

This in turn suggests a need to examine the possibility of moving toward a two-tiered army along the lines of the French Army. The French Army is essentially two armies in one: the first, the larger, consisting of less well-trained, short-term draftees, who by law cannot be compelled to serve in combat overseas (a legacy of the Algerian War), and organized and structured for the defense of Europe and metropolitan France; and the other, essentially a small *corps d'élite*, consisting of long-service professionals organized into highly specialized small units dedicated to the defense of French security interests in Africa and elsewhere overseas. Like France, the United States has a big-war commitment in Europe and small-war requirements outside Europe. Unlike the French army, however, the U.S. Army has not allocated its short-service and long-service manpower in a fashion designed to optimize small-unit cohesion for each task.

Decentralized Command Authority

Small wars, particularly those waged in remote areas against unconventional opponents, and above all special operations, demand greater decentralization of command authority and small-unit initiative than do large wars decided by weight of materiel. The history of the U.S. military's performance since 1945, however, has been in no small measure a history of excessive and often debilitating interference in the planning and execution of military operations by civilian leaders *and* senior military authorities. The problem of the micromanagement of military operations lies as much within the U.S. military as between the military and civilian authority. Worse still has been heavy reliance on operational doctrines, such as the current Rogers Plan and that espoused in the pre–AirLand Battle version of FM 100-5, and on training procedures, such as canned field exercises, that encourage tactical rigidity and discourage initiative and risk-taking on the part of subordinate commanders. The wide operational and tactical latitude accorded to British commanders was an indispensable condition for Britain's quick victory in the Falklands War and stands in sharp contrast to the excessive centralization of command authority that characterized U.S. military operations in Vietnam.

At the heart of the problem lies a Joint Chiefs of Staff so constituted as virtually to guarantee the subordination of wartime operational considerations to the advancement of the peacetime bureaucratic interests of the various services. The "tank" is less a military institution than it is a political forum for the resolution of preeminently political disputes.

All of this suggests the need not only for new operational doctrines and greater emphasis on free-play exercises and other training procedures, but also for fundamental reform of the Joint Chiefs of Staff.

PROSPECTS

To its credit, the U.S. Army, more than the other services, has recognized—*and is acting upon its recognition*—that small, semiconventional and unconventional wars are the most likely of all conflict

158   PROSPECTS FOR A FUTURE SYNTHESIS

environments to engage U.S. military power in the future. Implicit in that recognition is an understanding that the army's heavy, general-purpose force divisions dedicated to Europe's defense are not suitable for much of the

> environment of the world in which we live and the nature of the conflicts we are seeing. Many are at the lower end of the spectrum, in what might be termed subconventional warfare, terrorism, guerrilla warfare . . . or modified conventional warfare. . . .[13]
>
> Army leadership is convinced, based on careful examination of studies which postulate the kind of world in which we will be living and the nature of the conflict we can expect to face, that an important need exists for highly trained, rapidly deployable light forces. The British action in the Falkland Islands, Israeli operations in Lebanon, and our recent success in Grenada confirm that credible forces do not always have to be heavy forces.[14]

Pursuant to this appreciation, the army is now in the process of creating two new, ultra-light, 10,000-man infantry divisions designed to operate in close terrain against unconventional foes; one of the divisions will be specialized for mountainous warfare, the other for cold-weather operations. A third such division, and possibly two more, will be formed through conversion of existing standard unmechanized infantry divisions. The army also is increasing its investment in existing subdivisional units, such as Ranger battalions, specialized for the peculiar operational environments of much of the Third World. All of these measures are to be accomplished within the framework of the army's currently authorized manpower ceiling and without disturbing the army's ten (heavy)-division commitment to Europe.

Accompanying the light-division initiatives has been a highly encouraging set of experiments, known collectively as the New Manning System, designed to promote small-unit cohesion by keeping the same officers, NCOs, and men in the same unit for at least the first term of enlisted service. Although the New Manning System antedated the light-division initiatives and is deemed by some to be of potentially army-wide applicability, any major improvement in small-unit cohesion will redound to the greater benefit of small-war forces, whose small-unit cohesion requirements are, as noted, more extensive than those of big-war forces.

Finally, the army, in its adoption of AirLand Battle in 1982, has adopted an operational doctrine that would seem tailor-made for the most likely future conflict environments. Although AirLand Battle purports to be applicable to U.S. military operations anywhere and of any intensity, its long-overdue emphasis on decentralized command, tactical flexibility, and initiative and risk-taking on the part of subordinate commanders is especially pertinent to the operational and tactical requirements of small wars against elusive opponents in faraway places. The doctrine is certainly more appropriate than the one it replaced (the FM 100-5 of 1976), which focused almost exclusively on the European theater of operations, emphasized a rigid administration of firepower as the key to success, and postulated conditions of linear combat and a clear distinction between the forward battle area and a comparatively quiescent rear area.

To be sure, none of these major ventures in reform—the light-division initiatives, the New Manning System, AirLand Battle—have yet been fully integrated into the army. All are still in progress, and all have been challenged directly on their putative merits and indirectly by vested bureaucratic interests inherently resistant to change. The new light divisions have been condemned as too light to fight for more than a couple of days, even against low-intensity opposition. Some have also criticized their intended employment concepts as dangerously conventional. The New Manning System remains in experimental infancy and is staunchly opposed by those who favor retention of the army's current, highly centralized manpower system. Moreover, as noted, the New Manning System, even in conjunction with unit replacement, might be unable to provide the desired levels of small-unit cohesion within the framework of the all-volunteer force. And while the adoption of AirLand Battle has occasioned little controversy, it is one thing to rewrite a field manual and quite another to instill its teachings into those charged with military operations. There is, finally, the long-term problem, over which the army has no control, of the all-volunteer force's uncertain future ability to attract and retain qualified people in numbers sufficient to meet the demands of small-war conflict environments.

To identify and examine these real and potential obstacles to the army's ability to deal effectively with likely future combat environments—and thoroughly examined they must be—is not, however, to lose sight of the fact that the army, for the first time since World War II, is paying the requisite attention to precisely the kinds of conflicts that have actually characterized the world since 1945, and are likely in the future to characterize it, rather than continuing to focus, sometimes almost exclusively, on preparation for the least likely conflict in the least likely place.

NOTES

1. The attempted invasion of Cuba in 1961 was planned and carried out entirely by the CIA and did not directly involve U.S. military forces.
2. Even in Korea and Vietnam, Chinese and North Vietnamese regular forces, respectively, were employed mainly in an irregular or unconventional manner.
3. U.S. air operations over North Vietnam encountered stiff air defenses but no serious air-to-air challenge by the North Vietnamese air force.
4. Even were the Reagan administration's Strategic Defense Initiative technically feasible, affordable, and unsusceptible to potential Soviet countermeasures, its full implementation would not be possible for at least twenty, and probably thirty, years.
5. The United States has never conducted combat operations in sub-Saharan Africa and is highly unlikely to do so in the future, given domestic political constraints and the absence of profound U.S. security interests in that region.
6. See, for example, Jeffrey Record, "The RDF: Is the Pentagon Kidding?" *The Washington Quarterly* (Summer 1981).
7. During the six decades separating the end of the Crimean War and the beginning of World War I, the British fought no war in Europe and only one major war outside Europe—the Boer War.
8. The open expanses of the Sinai Peninsula, the scene of many stunning large tank battles, are not characteristic of most of Southwest Asia. Southern Iran and the Arabian Peninsula are largely mountainous and lack all but the most rudimentary road networks. The limitations of mechanized forces in such terrain were amply demonstrated in Israeli armored operations on the Golan Heights in 1973 and in Lebanon in 1982.
9. A notable exception to date has been high-performance tactical combat aircraft, whose transfer to Nicaragua may continue to be effectively deterred by a declared U.S. willingness to undertake unspecified military action in response.
10. As noted, it is assumed that U.S. ground combat forces and nuclear weapons will continue to be deployed in Korea and that their presence will continue to deter major conflict there.
11. Charles Ardant du Picq, *Battle Studies* (New York: Macmillan, 1921), p. 96.

12. See, for example, Martin van Creveld, *Fighting Power, German and U.S. Army Performance, 1939-1945* (Westport, Conn.: Greenwood Press, 1982).
13. Testimony of General John A. Wickham, Jr., U.S. Congress, Senate, Committee on Armed Services, *Hearings on the Department of Defense Authorization for Appropriations for Fiscal Year 1985*, Part 2. 98th Cong., 2nd sess., 1984, p. 525.
14. General John A. Wickham, Jr., to the Soldiers and Civilians of the United States Army, reprinted in *Army Times*, 7 May 1984, p. 10.

9 HUMAN RESOURCES AND MILITARY REQUIREMENTS
Strategic Considerations on Past and Future

Irving Louis Horowitz

A review of the literature on military resources reveals that the subject is most frequently dealt with—and, one is tempted to add, restricted to—logistical considerations. This chapter will show that the relationship between military manpower and strategy is a different and more complex kettle of fish. Human resource issues are built into the very definition of *logistics*, that branch of military science dealing with the procurement, maintenance, and movement of equipment, supplies, and personnel. Definitions of *strategy* are linked to the science or art of planning and directing military movements and operations. In other words, strategies presuppose logistics. As a result, conventional military strategy often takes for granted manpower considerations. Most writings on the subject of strategy do just that—take for granted that human resource requirements, no less than hardware components, have somehow been met as budgetary tasks.

Strategic aspects of manpower invariably involve three components, difficult to disaggregate, but equally vital: (1) the *relations* between contending forces and their relative strengths and weaknesses; (2) the *functions* of manpower, that is, whether they are offensively or defensively oriented, whether they are designed essentially for warfighting or peacekeeping purposes, and so forth; (3) the *goals* of forces, such as maintaining a global equilibrium or gaining specific victories with a minimal amount of risk. Relational, functional, and goal-oriented considerations together form a seamless whole. Seen in this way, desultory discussions of conscription versus voluntarism, or the quasi-economic uses of recruitment as a method of ameliorating youth discontent, can be placed in a larger, more meaningful societal context. It is instructive, therefore, in discussing the future of the military manpower environment, to look first to historical deliberations on the subject.

THE HISTORICAL RECORD

The long-term secular history of military manpower strategies is directly connected to difficulties that were a direct outgrowth of the rapid expansion of the military sector. It is characteristic of the political heritage of democracy, no less than of the economic heritage of private sector market preeminence, to downplay the issue of human resource planning and upgrade the notion of free choice. No one has expressed this better than Eli Ginzberg, the doyen of manpower planning. In his classic and still unsurpassed work, *The Ineffective Soldier*, he noted the following:

> Ever since colonial days, our country has manifested a negativism towards, even a fear of, large standing forces. In peacetime the nation would tolerate only a small professional Army and Navy. In emergency, citizens were expected to answer the call of the colors. Until the Civil War, it had been hoped that volunteers would come forward in sufficient numbers to meet their country's need; but in that war, as later in World War I, it was necessary to have recourse to the draft. A small professional Army in peacetime, reliance on the draft in case of major emergency—these were two of three legs of our military manpower planning. The third was a strong reserve to be built up in time of peace so that if an emergency occurred the Army would have trained units ready and would also have the commissioned and noncommissioned officers required to train the draftees.[1]

This pattern of professionalism held until the 1930s. Then, in 1935, it became clear that the New Deal meant to implement a dramatic change in the public sector—private sector mix by embracing a new view of democracy, one based far less on free choice and more on a notion of common national defense. Legislation created organizations such as the Civilian Conservation Corps that were quasi-military in character. The army had responsibility for receiving recruits;

equipping, housing, feeding, and sending them to appropriate camps throughout the country; and then organizing youngsters into effective work battalions. One can truly say that this recently concluded phase of military manpower planning lasted fifty years. Curiously enough, it began with a reconsideration of the public nature of civilian manpower-planning needs of a country gripped by depression at home and challenged by totalitarian movements abroad, and ended in 1981 due to similar domestic and overseas considerations.

Anticipating human resource requirements is an issue in planning. The question of military manpower needs is fundamentally the concern of military planning. To illustrate this, let us examine the first stage of the New Deal era, fifty years ago. Manpower issues were part of a liberal, even radical, thrust into the world of public sector planning. David Lilienthal articulated this principle in bold, visionary strokes relating to the Tennessee Valley Authority.

> In the last analysis, in democratic planning it is human beings we are concerned with. Unless plans show an understanding and recognition of the aspirations of men and women, they will fail. Those who lack human understanding and cannot share the emotions of men can hardly forward the objectives of realistic planning. . . . A great Plan, a moral and indeed a religious purpose, deep and fundamental, is democracy's answer to our homegrown would-be dictators and foreign anti–democrats alike.[2]

The same theme of manpower planning within a democratic context was struck in 1935 by John Dewey in the realm of ideology. Dewey viewed public planning as *the* necessary corrective to personal private advantage and avarice. For Dewey, a social order without planning could never be counted on to realize the highest aspirations of a democratic civilization, for it could never achieve a common set of core values.

> Organized social planning, put into effect for the creation of an order in which industry and finance are socially directed, is now the sole method of social action by which liberalism can realize its professed aims.[3]

It was the conservative reaction to this fusion of democracy and policy that led the charge against manpower planning. The entire New Deal came to be seen by conservative critics as a watered-down version of authoritarian modalities. Walter Lippmann sounded the clarion call against resorting to a totalitarian temptation such as manpower planning under other than the direst of circumstances.

> It is plain enough that a dictated collectivism is necessary if a nation is to exert its maximum military power: very evidently its capital and labor must not be wasted on the making of luxuries; it can tolerate no effective dissent or admit that men have any right to the pursuit of private happiness. No one can dispute that. The waging of war must be authoritarian and collectivist. The question we must now consider is whether a system which is essential to the conduct of war can be adapted to the civilian ideal of peace and plenty.[4]

Lippmann's free-choice question is the one we must ultimately take up and resolve as a society. But for now, it is sufficient to note that human resource planning—the determination of needs in terms of broad socio-political goals—is hardly part of the folklore of Republican conservatism. Rather, it is historically linked to the folklore of Democratic liberalism. Yet, what took place a quarter century later in the context of Vietnam revealed that the frontal assault against manpower planning most often came from liberal and radical sources of American society. This might come as something of a shock to those reared within an earlier New Deal mind-set. It should caution all of us to beware the bland, abstract notion that manpower planning is linked to any single ideology or political party.

In American terms, at least, planning has been part of an overall strategy to solve problems. During the depression, this meant alleviating the problem of mass civilian unemployment by means of a conservation corps. During World War II, the Korean War, and finally Vietnam, it meant generating the troop strength necessary to neutralize enemies, if not win those wars. It is sobering to remember, nonetheless, that broad manpower management won neither the internal war against poverty nor the external war against totalitarian enemies. This should serve as a caution against using a single-variable formula for discussing broad manpower strategies.

Human resource concerns have been discussed for so long in the context of military conscription that the novice may well be excused for thinking that this issue has been thoroughly resolved, at least in terms of peacetime conditions. But the concept of an all-volunteer armed force, once institutionalized, stimulated only further debate. The work in the late 1960s of the National Advisory Commission on Selective Service, and an earlier Ford Foundation-sponsored conference held at the University of Chicago, provided the major focus for intellectual debate on the issue well until the early 1980s.[5] In the mid-1960s the liberal tradition, previously supportive of military planning, reversed its position. Anthropologist Sol Tax, in his introduction of the edited volume, *The Draft*, provides a keen sense of the situation at that time:

> The conference itself was an exciting event, with "We won't go!" student meetings outside and a variety of political machinations inside. The chairman often had difficult moments in keeping down attempts to turn the educational discussion into a parliamentary body.... The conference threatened to split apart the last evening, when arrangements were made for a nationwide telecast of the next morning's concluding session; the issue was never settled because a local accident involving the fall of an elevated train from its trestle suddenly removed the television cameras to that scene.[6]

This is neither the time nor the place to review the antiwar literature spawned by the war in Vietnam, except to note that large-scale philosophical issues were involved: free choice versus common defense; social responsibility versus individual rights; top-down versus bottom-up selection and recruitment patterns; civilian manpower needs versus military manpower needs, and so forth. Issues became ideologically criss-crossed: Advocates of the poor felt that conscription placed too great a burden on the lowest rung of society, while creating an evasion mentality among middle-class youth. Others argued that the all-volunteer force might get recruits, but not of the quality needed for its advanced technological requirements.

Current discussion, however, takes place in a markedly changed national climate. Even if global alignments have remained essentially unchanged over the past several decades, the emergence of a renewed national consensus combined with a belief in free market principles will dramatically alter the quality, as well as the terms, of the manpower debate.

THE LESSONS OF VIETNAM

Human resource considerations can be viewed in capital-intensive or labor-intensive terms, but they cannot be taken out of either context without severe distortion. The determination of human resource requirements in the armed forces is a direct function of the type of conflict for which the military is being readied. If the anticipated next phase of armed conflict is seen as taking place in a guerrilla environment, then the number of troops required will be immense. Third World armed conflict requires high manpower needs: Fundamental verities of training, courage, morality, and purpose remain central. Anticipated struggles between advanced industrial powers, or if you will, between the First and Second worlds, require capital-intensive scenarios. "Star wars" scenarios involving offensive and defensive capabilities at the nuclear level would presumably require less manpower but far greater technical expertise in computer design and programming than that required in past conflicts.

The Vietnam War, whatever one concludes is its ultimate meaning, entailed a classic case of mixed economic metaphors. The United States engaged in a conflict with both a guerrilla and a national opponent, fighting an unconventional war with the South and a quite conventional war with the North. As a result, U.S. strategy was at one and the same time based on large numbers of soldiers and on an even larger amount of conventional hardware. There was, in short, an effort to match military troops against guerrillas at parity ratios. But as demands for withdrawal and cutbacks increased, American military strategy turned capital intensive, focusing on large numbers of search and destroy missions, an increased use of helicopters, long-range bombings, village pacification programs—anything that would alleviate the need for continuing manpower increases. In short, it was not an abstract formula of manpower to weaponry, but the relationship, the fit, or both, of manpower to weaponry in terms of the specific struggle underway or anticipated, that was used to resolve strategic manpower considerations.

The extent to which manpower size is maintained is, in part, a reflection of commitment. Great powers are not easily enticed into making all-out efforts when the stakes are limited. But for the "other side" the perception is frequently of unlimited stakes. Eliot A. Cohen, in his analysis of constraints on America's conduct of small wars, parenthetically puts his finger on constraints on manpower *and* firepower deployment, namely, the unwillingness to take extreme risks when perceived rewards are limited.

> These two kinds of wars require very different kinds of military forces and systems of military command, for it is not the case that an army suited for European war can handle all other contingencies with aplomb.... Small war is not a "half" a war, but rather a completely different kind of conflict. It takes its peculiar coloration from the geopolitical circumstances which call it forth, and hence requires special means for its conduct. In order to wage small war successfully, a military establishment must meet its requirements in five respects: expectations (vis-à-vis the foreign and domestic political context of such conflicts), doctrine, manpower, equipment, and organization.

In all respects the American defense establishment—civilian as well as military—is deficient. In some cases, what is lacking is an understanding of the problem; in others, the ability to implement solutions.[7]

While Cohen's remark does not focus exclusively on the Vietnam War, it is evident that this notion of uneven stakes is particularly applicable to that conflict and other similar conflicts in which one party throws everything into battle and the other, the American side, counts every injury. The notion of uneven stakes reflects the limits of manpower no less than the limits of firepower, in subnuclear wars no less than in anticipated nuclear wars.

While there is no longer any doubt that a stronger and more cohesive force in South Vietnam could have produced a different outcome and that the breakdown of fighting morale among American armed forces and their leadership played a significant role in that defeat, there is also no question that South Vietnam fell as a result of an onslaught conducted with heavy conventional weapons, not in a guerrilla conflict.[8] The guerrilla phase led to a protracted war. However, the question of victory and defeat was fueled by the manpower and material strength of the North Vietnamese.

By far the most persuasive statement on Vietnam as it bears on manpower considerations was made by John M. Gates. He pointed out the fallacy of blaming significant military errors on academic advisors or civilian politicians.[9] The pleasant fiction that absolves the military of all responsibility for the Vietnam debacle is far more difficult to sustain if one takes into account the unconventional nature of that war. Guenter Lewy further bears out this position in noting:

> An analysis that denies the important revolutionary dimension of the Vietnam conflict is misleading, leaving the American people, their leaders, and their professionals inadequately prepared to deal with similar problems in the future. The argument that faulty strategic assessment and poorly articulated goals doomed the American military to faulty operations in Vietnam only encourages military officers to avoid the kind of full-scale reassessment that failures such as that in Southeast Asia ought to stimulate. Instead of forcing the military to come to grips with the problems of revolutionary warfare, such analysis leads officers back into the conventional war model that provided so little preparation for solving the problems faced in Indochina by the French, the Americans, and their Vietnamese allies. Such a business-as-usual approach is much too complacent in a world plagued by the unconventional warfare associated with revolution and attempts to counter it.[10]

What this suggests in terms of manpower is the need not only for synergistic balance with technology but for symmetrical balance between conventional and guerrilla strategy and tactics as well. The quality and the type of manpower recruited is thus as significant a portion of the strategic equation as are the numbers of people involved.

Human resource issues have been often neglected because of the propensity of civil and military authorities alike to emphasize hardware issues, specifically nuclear weaponry. Without minimizing the significance of nuclear arsenals, stockpiles, innovations, and the like, the current emphasis on Star Wars is a capital-intensive approach that may lead to the neglect, benign or otherwise, of the labor-intensive, or manpower, aspects of military might.

It may well be the case that manpower strategies cannot by themselves serve to determine larger political or military parameters. That said, it is equally important to assert the reverse: that new technology will be no magic solution to manpower requirements. Martin van Creveld, in his work on *Command in War*, puts the capital-intensive formula into a balanced framework.

> Far from determining the essence of command, ... communications and information processing technology merely constitutes one part of the general environment in which command operates. To allow that part to dictate the structure and functioning of command systems, as is sometimes done, is not merely to become the slave of technology but also to lose sight of what command is all about. Furthermore, since any technology is by definition subject to limitations, historical advances in command have often resulted less from any technological superiority that one side had over the other than from the ability to recognize those limitations and to discover ways—improvements in training, doctrine, and organization—of going around them. Instead of confining one's actions to what available technology can do, the point of the exercise is precisely to understand what it cannot do and then proceed to do it nevertheless.[11]

In short, the relationship between manpower and hardware, between labor-intensive and capital-intensive elements in the overall military picture, holds the policy-relevant answers.

HUMAN VERSUS MATERIAL RESOURCES

Human resource needs stand as the quintessential expression of civil-military relations. This is so not only in the prosaic sense that selective service is a process by which youths move from civilian to military sectors. But in the more profound sense, military human resource needs, whether satisfied through voluntaristic or conscriptive mechanisms, presume the right of the nation-state to determine the contexts in which youths can be removed from homes and families as well as the conditions under which the sacrifice of lives is deemed warranted. In an America increasingly described in metaphors of narcissism—something quite beyond free choice—manpower requirements stand as a constant reminder, often an irritant, that some sort of implicit social contract between state and individual for the common defense remains intact. If the relationship of society to the individual tends to be cast in terms of benevolence and rights, those of the state and the individual clearly fall into the realm of obligations and duties—a much harsher and traditional rhetoric by any American standard.

Common wisdom in the post-Vietnam era indicates that "the prestige, self-image, and material health of military institutions will prosper if the military can convince civilians and themselves that wars can be short, decisive, and socially beneficial."[12] This is a tall order under optimal circumstances, but in a post–nuclear environment characterized by guerrilla insurgency, terrorist assaults, and ambiguous battlefield environments, it makes the neo-Hobbesian goal of short, nasty, brutish (not to mention decisive and beneficial) wars ever more difficult to achieve. When current mores about both the high value of individual life and nonmilitary solutions to political problems are factored into this equation, the task of justifying large manpower needs grows much more complex. Indeed, in such social and psychological circumstances, it is far simpler to opt in favor of technological increases and manpower decreases.

The relation between material production and manpower planning is, however, not mechanistic. The armed forces cannot simply substitute sophisticated hardware to displace manpower or allocate budgetary funds for a Star Wars scenario as a mechanism for holding manpower needs constant. In one of his more stunning observations, Michael Handel has pointed out that "the United States, a modern and advanced industrial nation with armed forces possessing the most sophisticated weapons in existence, has gradually been reduced to recruiting soldiers on the level of an underdeveloped society."[13] Clearly, the continued, and in some instances increasing, gap between the quality of American weapon systems and the lack of quality of military personnel available to utilize such refined hardware constitutes a major problem in manpower deployment and planning.

The resolution of this dilemma cannot be one sided. At the prima facie level, manpower personnel must be considerably upscaled. This can be effectively accomplished through the usual mechanisms of increased pay, greater choice of service branches, and shorter terms of enlistment and re-enlistment. But at the other pole, there must also be a radical shift in weapons design strategies. The need for simplification as well as modernization becomes imperative in the light of the quality of human resource availabilities. Battlefield experience, or at least simulated conditions of combat, must be brought to bear on the research and development of military hardware. Technical and engineering personnel think in terms of state-of-the-art instruments and weapons. But increasingly, the armed forces need intermediate as well as advanced technology, based on limited manpower capacities and capabilities. This means simplification and rationalization of tactical aircraft, cheaper night-sight equipment, and trucks and tanks that can be driven as easily as autos. As long as current world conflicts remain low intensity rather than nuclear in nature, weapons and manpower needs must continue to address the conventional present, even as it evolves into a star-wars future.

Human resource strategies are part and parcel of national planning. Because manpower requirements for the armed forces are directly determined by the federal budget, they become subject to allocation mechanisms that extend beyond those of the private sector marketplace. Such manpower requirements are set not simply by direct military needs but also by the anticipated needs of the larger society. Military manpower can thus provide potential leverage on basic fluctuations in the economic system of free enterprise. It can and does provide a cushion against undue youth employment during economically difficult times and must compete against private sector inducements when the economy is particularly strong.

Human resource strategies must be responsive to both direct military needs and indirect economic pressures and counterpressures. The question of resources is always one of quality as well as quantity. Hence, effective measurements require a widespread effort to include social indicators as well as numbers. In wartime conditions, when conscription is acceptable, normal fluctuations are done away with. But increasingly, even under quasi-wartime conditions, such as the Korean War and the Vietnam War, where mobilization is fragmentary and partial, the competitive advantage stays with the civilian economic sector. Indeed, since during wartime the dangers to life and limb increase, production mobilization also increases. Thus, competition is conditioned not simply by economic or market fluctuations but by social norms. The advantage of military manpower planning is the commanding position of the national system to achieve its resource requirements through budgetary allocations. The weakness of such planning is the ubiquitous nature of the economic marketplace at a particular time with respect to unanticipated conflicts or unintended outcomes.

In sub-nuclear military conflicts the need for manpower is likely to increase in the short run, while the capacity to meet these needs is likely to decrease. Despite the temporary advantages of more enlightened and better qualified troops—for reasons such as a renewal of patriotic fervor, a higher conservative strain in the youth of the mid-1980s, and rapid response to Soviet incursions and aggressions—in long-range terms, maintaining present personnel levels of 2.3 million will be difficult on an all-volunteer basis. As Charles Moskos and Peter Braestrup recently pointed out:

> Just to maintain the current strength in 1986-1993 of the active and organized reserve forces (a total of three million men), the military will need to enlist one out of three eligible males—"eligible" meaning able to meet current physical and mental standards. If all college youth are excluded, one out of two eligibles will have to be recruited. Inevitably, the question of revising the draft—presumably a two-year draft by lottery with low pay but with some sort of G.I. Bill—will come again shortly.[14]

While demographics argue in favor of some sort of service requirement for young American men and women, neither the executive nor legislative branches of government, nor for that matter either of the two political parties in this country, sees this as a priority agenda item. In the absence of an overt, direct threat to American vital interests, the prospects for renewal of a draft remain weak; hence, the capacity to enlarge the armed forces remains limited. The drive for technological solutions to manpower problems is fueled not only by military preferences for a deterrence environment but by a political impasse that necessitates avoiding protracted conflicts of a conventional sort at all costs.

This caveat registered, the struggle over the all-volunteer force is hardly closed. For while there is a general consensus that the raw numbers of people serving in the reserves and on active duty are adequate and that indicators of manpower readiness are appropriate, the problem of quality remains on the military agenda. Lawrence J. Korb, the Pentagon manpower chief, places the matter in unvarnished numerical terms.

> [W]e will never get the people we need to carry out the job. Whether they are civilian employees, active duty military, or reserve military, the people that we need to do the job are not there. Last year we asked Congress for what I thought was a relatively modest increase in the active forces of 40,000 people to man new equipment which had already been authorized. We received approximately 10,000 men. This year we scaled down our expectations, sensing that the Congress would say they felt we had too many people, we asked for 30,000; we will be lucky if we get 15,000. What we really needed over the past 2 years was about 100,000 more people on active duty. What we ended up netting is about 25,000 people, similarly with civilians.[15]

In the mid-1980s the terms, no less than the players, have shifted dramatically. Democratic Party leaders such as Senator Gary Hart and Representative Robert G. Torricelli have drafted legislation to establish a commission to study the possibility of requiring all young people upon graduation from high school to do national service, ranging from military conscription to social work for the elderly. But it is now the Republicans who have become politically wary of such mandated service. Secretary of Defense Caspar W. Weinberger sees the all-volunteer military as attracting quality personnel; hence he is "constitutionally opposed to fixing something that is not broken." General John W. Vessey, Jr., chairman of the Joint Chiefs of Staff, added that "universal service may have a value to the country, but its value is not to the Department of Defense."[16] The end of the draft in 1973 essentially brought to an historic conclusion the period of treating military manpower needs differently than civilian manpower requirements. Since 1973, and in the absence of an emergency, economic incentives coupled with technological displacements have become the operational code of military manpower fulfillment. Military service has been redefined as civic education.[17]

Foreign policy is a special variant of domestic policy applied to overseas contexts; by the same token, domestic policy is a form of foreign policy, insofar as regional, ethnic, racial, class, gender, and age differences, among other considerations, enter into decisions and programs intended to move masses and elites toward certain articulated goals.[18] It might well be claimed that manpower allocation decisions in general exemplify this cross-fertilization of foreign and domestic policy. It might further be said that racial issues are specific illustrations of a fusion of manpower policymaking.

The issue of racial composition in the all-volunteer force has been widely discussed. It probably is the most heatedly examined issue, since the Vietnam War made plain the enormous number of black troops involved in combat duty. Indeed, one out of every twelve black male workers between eighteen and twenty-one years old is in the military, making the armed forces a major source of employment and income for black Americans.[19] It is fully known that military service offers blacks better opportunities for responsible work at fair compensation than are available to minorities in many sectors of private enterprise. One can readily see how foreign policy and domestic policy considerations converge in any calculation of the impact of black Americans on the all-volunteer military.

The emergence of serious issues in this environment, one in which minorities comprise 26 percent of our 2.1 million-member active duty force, has been made plain by researchers. As the roster of technical jobs grows, the pool of manpower capable of adequately handling new levels of technical responsibility diminishes. *The Profile of American Youth* report showed that while 78 percent of white youths score in the top three categories, only 28 percent of black youths (and 41 percent of Hispanic youths) do as well. Should the armed services recruit only from these top categories, 72 percent of black youth would be excluded. This, in turn, would seriously affect current black youth employment opportunities.[20]

The problem is that recommendations on this aspect of human resource management are often at cross-purposes: spend more money on compensation and recruitment; turn over more military jobs to civilians; increase the role and number of women in an effort to get top-quality personnel; lower entry standards. While in the abstract each of these approaches is feasible, the interpenetration of foreign and domestic policies changes the vectors of actual power relationships. The issue of manpower strategy, therefore, must become increasingly sensitive to both domestic and foreign policy considerations. While a broad (if thin) consensus apparently exists that the all-volunteer military would remain the basic policy posture, at least for the balance of the decade, dealing with the problems connected with this approach is becoming ever more paramount. Higher pay scales for military personnel bring about demands from legislative purse string managers for higher performance levels. Budgetary squeezes bring about demands for lower numbers of military personnel. In short, as the armed forces become recognized as a permanent part of the economic sector, domestic considerations, no less (and sometimes more) than raw strategic-military concerns, will prevail.

THE U.S.-SOVIET BALANCE

We come now to the final point in the analysis of human resource strategies: the conduct of conflict by symbolic means, that is, an outline of manpower circumstances that, like hardware structures, serve to deter aggression. In this epoch, we are referring to the specific relationship that obtains between Soviet and American forces. The noteworthy efforts of Edward N. Luttwak to appraise comparative numerical strengths of Soviet and American manpower capabilities deserves serious attention. The burden of Luttwak's remarks rests not with Soviet parities but with American disparities. Soviet ground forces create strategic imbalances in every possible war zone.

> The gross totals are well known, and mean little. As against the 30 large divisions of the U.S. army and marine corps, active and reserve, the Soviet army has 194 divisions, smaller by a third on average but just as heavily armed. One-third are fully manned, one-third are half and half, and the rest are mostly manned by reservists—but all Soviet divisions are fully equipped, even if not with the latest and best, and all have a full-time professional cadre, even when their line units are manned by reservists.... In the five possible war theaters of the North Atlantic alliance—northern Norway, the "central" front in Germany, northeast Italy, the Thrace frontier of Greece and Turkey with Bulgaria, and the Turkish border with the Soviet Union in remote eastern Anatolia—it is clear that the ground forces of both the United States and its allies, those already deployed in peacetime and those to be mobilized, would be outnumbered, outgunned, or both.

After reviewing other areas of potential strife between the superpowers, Luttwak concludes on a sobering note:

> Thus on every possible major front we encounter the powerful arithmetic of the Soviet army. By integrating reserves with active units and providing full equipment, the Soviet army is a very effective producer of armor-mechanized divisions. Not at all suited for overseas expeditions, dependent on rail trans-

port for large movements between the different fronts separated by several thousand miles, these divisions are nevertheless powerful instruments of offensive war wherever the Soviet Union may seek to enlarge its empire.[21]

The human resource factor, far from being peripheral, is central to strategic-military concerns. Nuclear parity, as an area unto itself, leaves untouched the enormous conventional superiority of the Soviet Union over the United States. Furthermore, since the nuclear threshold of which Herman Kahn spoke has not been crossed since Hiroshima and Nagasaki, the importance in strategic terms of low-intensity conflict considerations grows, rather than diminishes, over time. In the sorts of wars that are routinely being fought today—guerrilla conflicts—the ratio of conventional forces to insurgency forces must be quite high, ranging in some estimates from three to one and in others up to ten to one, necessary simply to maintain a standoff. Thus, even if we ignore the quality of manpower, confining our discussion to quantity, it is clear that the present state of benign neglect only serves to enhance unrealistic expectations of advanced hardware developments.

The level of technological parity between the United States and the Soviet Union is much greater than that usually reflected in literature reviews. The Soviet Union has moved a long way from its post-World War II posture of *substituting* manpower quantity for hardware quality. That posture made sense in an earlier period of profound nuclear imbalance. It no longer characterizes the Soviet posture. While it remains correct that the qualitative edge in hardware remains with the United States, and will remain so as long as the United States maintains superiority in the computer and electronic industries, the U.S.-Soviet gap in these areas has narrowed considerably. While the range of the new technology in the Soviet Union has hardly penetrated the civilian sectors of Russian life, it has gone far and deeply into military life. In areas ranging from space exploration to the production of tanks and armored personnel carriers, the Soviet Union already matches or has improved upon U.S. levels. Furthermore, Soviet technology remains substantially cheaper, in key aircraft areas by as much as one half to two thirds of U.S. equivalents.[22]

The Soviets make weapons changes on the basis of battle-proven capacities, while the United States makes alterations on the basis of industrial capabilities—often on the basis of prototype weapons generated in technological vacuums, isolated from manpower capabilities or considerations. While Soviet manpower strength is greater than that of the United States, the ratio (roughly two to one) has remained constant for some time, at least two decades. It is probably the case that in terms of technological quality, no less than manpower quantity, the advantage at present goes to the Soviet Union. It might well be that the Soviet model provides a strategic ideal toward which the American model should move; a more highly integrated relationship between manpower and hardware, entailing a simplification of the latter and an expansion of the former.

While it may be correct that in nuclear exchanges, or for that matter in nuclear technology, the United States maintains a slight edge, this tactical advantage has little meaning and even vanishes with respect to conventional, low-intensity conflicts. At such conventional levels, the crudities of Soviet hardware, frequently disparaged, and the overwhelming Soviet manpower superiorities, rarely disparaged, come into play. In a sobering article on conventional force deterrence, James M. Garrett has noted the strategic impact of this situation.

> NATO's numerical inferiority and the [Warsaw] Pact's ability to mass overwhelming forces in several sectors require the Alliance to practice strict economy of force in two senses: to commit only the forces necessary for a given mission and to avoid attrition except when a very favorable ratio of losses can be expected. NATO cannot afford to exchange losses with the Pact in a one-for-one ratio. Compared to the Pact, the [Western] Alliance must keep a higher proportion of its troops in action and, therefore, cannot afford to have substantial numbers bypassed or otherwise out of action.[23]

Like many other military analysts, Garrett operates under the assumption that the present balance of military forces will remain intact. Hence, he speaks of the need for better small-unit training, more adequate training for individual soldiers operating in conditions of relative isolation, and increased tactical intelligence capabilities. While these are clearly measured responses, they do not address the extraordinary and continued disparity in manpower allocations of the U.S. and Soviet militaries. Keeping in mind Soviet technological advances at low-intensity levels, it is difficult to see how pedestrian appeals to better training or better infra-red or radar systems can conceivably change battle outcomes. At some level, the question of manpower, like that of firepower, becomes a concern with political as well as military implications.

Given the internal political realities of the United States and the reticence of America's allies to move beyond a force-deterrence position in Europe, artificial and sudden increases in manpower are unlikely to happen or, if slight changes do take place, unlikely to change events dramatically. A more likely change is in the global structure of alliances. For many years since the end of World War II, most defense planning and strategy has been almost totally dominated by considerations of the European theatre, the area in which the two major superpowers face each other. But actual conflicts have been located in the Third World, not in Europe. Warfare in Asia, Africa, and Central America has been pandemic, yet strategic discussion in the United States continues to be dominated by European strategies and strategists.[24]

The changed world posture of China, with its extraordinary 900 million plus population, becomes central to manpower discussions of any new Western alliance. Admittedly, the contribution of China to a Western technological shield is slight, but its potential and actual contributions to a Western shield in places like Vietnam, Cambodia, and even potentially in India cannot be entirely discounted. If India moves toward a Western political posture, then the full weight of manpower concerns will be decisively tipped. Too much human resource analysis has focused exclusively on Europe. Hence, the global-political aspects of human resources have been overshadowed by domestic constraints and restraints, Military manpower concerns are intimately linked to demographic patterns; hence, the profound shift of China and, lately, of India should, in the years immediately ahead, change manpower scales considerably, restoring a degree of symmetry between East and West.

CONCLUSION

If the future is in some ways an extrapolation from the present, it is clear that human resource considerations will not dissolve in some magic post-industrial policy formulas in which material production will replace or displace human manpower. Quite the contrary, one Vietnam legacy is a realization that human resource concerns remain high in unconventional military conflicts. Precisely in the context of sub-nuclear conflicts (Vietnam for the United States, Afghanistan for the Soviet Union) the limits of military hardware become painfully apparent.

The specific characteristics of human resources for the balance of the decade, all conflictual events being equal, will include the following:

1. Closer integration of hardware development and research with human resource talent and needs.
2. Closer integration of military manpower with overall economic development and planning, especially the utilization of military manpower as a cushion to even out economic fluctuations.
3. Consideration of human resource demands as largely political issues. The question of an all-volunteer versus a conscription military raises political questions; opposition from both the left and the right inhibits restoration of a draft under peacetime conditions.
4. Continued disparities between American and Soviet demographic potentials and an effort to reduce strategic disparities by means of technological breakthroughs and nuclear-deterrent scenarios.
5. Actual combat outcomes determined by changing relationships of forces in global political terms rather than by artificial or sudden increases in manpower deployments.

Direct policy intervention is only one consideration that should inform decisions regarding future human resource requirements. Ubiquitous political alignments and broad patterns of economic growth are other, perhaps much more significant considerations. We have gone through two broad periods of military manpower approaches: the period from 1789 to 1929, in which the size of standing armed forces was determined by emergency situations, followed by the 1930 to 1980 period in which general economic concerns coupled with global political interests determined the size and character of armed forces.

The new third stage in human resources, one we are entering, is beset by a major contradiction: an *economic* philosophy of government that emphasizes private sector initiatives and innovations, coupled with a *political* philosophy based on the confrontation of democratic and totalitarian values, in which the public sector is best exemplified by a talented, sizable, and professional (albeit voluntary) armed force. How the government orchestrates the private economy (based on self-regulation) with the public economy (based on government regulation) will likely determine the course of human resource planning and policy in the military for the balance of the twentieth century.

NOTES

1. Eli Ginzberg, *The Lost Divisions*, vol. 1 of *The Ineffective Soldier* (New York: Columbia University Press, 1959), 20-21.
2. David E. Lilienthal, *TVA: Democracy on the March* (New York: Harper Bros. 1944), pp. 120-211.
3. John Dewey, *Liberalism and Social Action* (1935; reprint, New York: Capricorn Books, 1963), p. 54.
4. Walter Lippmann, "Planning in an Economy of Abundance," *The Atlantic Monthly*, January 1937, pp. 39-46.
5. *Report of National Advisory Commission on Selective Service*, commission headed by Burke Marshall (Washington, D.C.: U.S. Government Printing Office, 1967).
6. Sol Tax, ed., *The Draft: A Handbook of Facts and Alternatives* (Chicago and London: The University of Chicago Press, 1967).
7. Eliot A. Cohen, "Constraints on America's Conduct of Small Wars," *International Security* 9, no. 2 (Fall 1984): 166-67; also see Andrew J.R. Mack, "Why Big Nations Lose Small Wars: The Politics of Asymmetric Conflict," in *Power, Strategy and Security*, ed. Klaus Knorr (Princeton, N.J.: Princeton University Press, 1983), pp. 126-51.
8. Harry G. Summers, Jr., "Vietnam as Unending Trauma: Defense Without Purpose," *Transaction/SOCIETY* 21, no. 1 (November/December 1983): 4-17.
9. John M. Gates, "Vietnam: The Debate Goes On," *Parameters: Journal of the U.S. Army War College* 14, no. 1 (Spring 1984): 15-25.
10. Guenter Lewy, "Some Political-Military Lessons of the Vietnam War," *Parameters: Journal of the U.S. Army War College* 14, no. 1 (Spring 1984): 2-14.
11. Martin van Creveld, *Command in War* (Cambridge and London: Harvard University Press, 1985), p. 275.
12. Jack Snyder, "Civil-Military Relations and the Cult of the Offensive, 1914 and 1984," *International Security* 9, no. 1 (Summer 1984): 121-22.
13. Michael Handel, "Numbers Do Count: The Question of Quality versus Quantity," *The Journal of Strategic Studies* 4, no. 3 (September 1981): 225-60, especially pp. 245-46.
14. Charles C. Moskos, Jr., and Peter Braestrup, "American's National Security: The Human Element," *The Wilson Quarterly* 7, no. 5 (Winter 1983): 131-33. See also Michael Useem et al., "The Rise and Fall of the Volunteer Army," *Transaction/SOCIETY* 18, no. 3 (March/April 1981): 28-60.

15. Lawrence J. Korb, "The Job is Half Done: Keynote Address: America's Defense Picture," *Defense Transportation Journal* (December 1984): 20-21.
16. "No Need Seen for Universal Service," *The Washington Post*, 22 February 1985, p. A7.
17. Morris Janowitz, *The Reconstruction of Patriotism: Education for Civic Consciousness* (Chicago and London: The University of Chicago Press, 1983), pp. 43-72.
18. Irving Louis Horowitz, "Foreign Policy: Domestic Policy by Overseas Directives," in *Policy Studies Review Annual*, ed. Ray C. Rist (New Brunswick and Oxford: Transaction Publishers, 1985), pp. 259-63.
19. Martin Binkin, *America's Volunteer Military* (Washington, D.C.: The Brookings Institution Press, 1984), p. 35.
20. James R. Daugherty, "Minorities and Military Recruitment," *Focus: Joint Center for Political Studies* 13, no. 1 (January 1985): 3-4.
21. Edward N. Luttwak, "Delusions of Soviet Weakness," *Commentary* 79, no. 1 (January 1985): 32-38.
22. See John M. Collins, *U.S. Soviet-Military Balance: Concepts and Capabilities, 1960-1980* (New York: McGraw-Hill, 1980), esp. pp. 101-106; Tyrus W. Cobb, "Tactical Air Defense: A Soviet-U.S. Net Assessment," *Air University Review* no. 30 (March/April 1979): 18-39.
23. James M. Garrett, "Conventional Force Deterrence in the Presence of Theatre Nuclear Weapons," *Armed Forces & Society* 11, no. 1 (Fall 1984): 59-83.
24. Irving Louis Horowitz, *Beyond Empire and Revolution: Militarization and Consolidation in the Third World* (New York and London: Oxford University Press, 1982), pp. 74-87.

VI ROUNDTABLE DISCUSSIONS

10 MANPOWER AND STRATEGY
Issues in Methodology and Analysis

George W. Sinks

The issues addressed by any body of scholars or analysts, as well as the manner in which those issues are treated, tend to be dictated by the disciplinary boundaries established by the academic community and by the institutional boundaries erected by the bureaucracy. A roundtable discussion was convened, therefore, to assess the impact of various methodological and analytical considerations in the manpower-strategy linkage. Dr. Franklin Margiotta, President of Pergamon-Brassey's International Defense Publishers, chaired the roundtable. Roundtable participants included:

- Colonel Trevor N. Dupuy (U.S. Army, retired), President and Executive Director of the Historical Evaluation and Research Organization, who provided the historical perspective
- Dr. Mady W. Segal, Associate Professor of Sociology at the University of Maryland, who provided the sociology/social psychology perspective
- Dr. Richard Ned Lebow, Professor of Government and Director of the Peace Studies Program at Cornell University, who provided the political science perspective
- Dr. Richard V. L. Cooper, Partner-in-Charge of International Trade Services and the Economic Studies Group of Coopers & Lybrand, who provided the economics/econometrics perspective

- General Paul F. Gorman (U.S. Army, retired), former Commander in Chief of the U.S. Southern Command, who provided the practitioner's perspective

Dr. Margiotta opened the discussion with a brief review of the past relationship between manpower and strategy. Military manpower issues, he argued, all too often have been applied against strategic considerations through a series of filters. The linkages between the two remain weak and, in many cases, are not well understood. The study of manpower and strategy involves many disciplines—history, political science, sociology, pyschology, and economics, with economics often predominating. What may be required in the future is less economically oriented analysis and more balanced, multidisciplinary analysis.

HISTORICAL PERSPECTIVE

After stressing the fundamental importance of history to the study of manpower, Colonel Dupuy reviewed a number of ongoing historical studies that address contemporary manpower/strategy concerns. These studies include an analysis of the Soviet and German personnel replacement systems during World War II, a study of the relative performance of German and Allied reservists during the opening phases of World War I, and a review of the preparedness debate in the United States in 1914-1917. Additional subjects of research dealt with the role and function of U.S. home defense forces during World War I, World War II, and the Korean War; allied personnel replacement systems during World War I and World War II; and the partial mobilization of the U.S. economy and manpower base during the Korean War.

Dupuy then offered some general comments on manpower/strategy issues and on earlier presentations at the conference. First, he differed with the concept of power put forward in an earlier session. In brief, he argued that will and intent are part of a nation's specific policy, not part of an abstract notion of power. Second, he disagreed with John Keegan's assertion that Clausewitz was an eighteenth-century thinker in terms of his views on maneuver and mass. Dupuy also found fault with Keegan's argument that Americans learned from the European military experience of the twentieth century the folly of raising mass citizen-based armies. Dupuy pointed out that the United States had, in fact, done just that during both World War I and World War II; the key difference was that the American mass armies did not suffer the horrendous casualties of their European counterparts. Moving to broader issues, he argued that the attrition rates of wartime equipment constitute a serious problem that deserves further research and analysis. He also asserted that the concept, if not the reality, of the universal military obligation is enshrined in American history. The idea of conscription is thus not alien to this nation's experience, contrary to the claims of some observers.

SOCIOLOGY/SOCIAL PSYCHOLOGY PERSPECTIVE

At the outset of her presentation, Dr. Segal identified four levels of sociological analysis—societal, organizational, group, and individual. At the societal level, the focus of analysis is on the relationship between the military and society, or put another way, on civil-military relations. Key concepts and considerations include the idea of the citizen-soldier, the requirements of citizenship, the obligation of military service, and the representativeness of the armed forces. Also important at this level are the public's view of the military and of military service, youth attitudes toward military service in general and the draft in particular, and the impact of demographic trends (e.g., a declining manpower pool) on military manpower policies. Two basic issues emerge at the societal level: the legitimacy of the military in society and the relationship between deterrence and large standing forces.

At the organizational level, the analytic focus is on trends in military organization. Key issues include the extent of convergence and divergence between civilian and military organizations, the development of professionalism within the military itself, the requirements of good leadership, and the causes and consequences of careerism—both in the enlisted ranks and the officer corps. Segal argued that most Americans do not want a mercenary military force. Thus, too much emphasis on financial issues degrades the legitimacy of the military institution in the eyes of those who fund and support it.

The third level of analysis identified by Segal concerned military group dynamics. Important issues at this level include unit cohesion and morale, the anatomy of military effectiveness, and the relationship between unit cohesion and success in combat. Segal also noted increased interest by sociologists in such issues as the conflict between the military family and the military institution, the effect of high technology on small-unit cohesion, and the differing impact of crew-served versus individual weapon systems.

At the individual level of analysis, several issues are important. They include the individual's motivation to enlist in the military, military socialization, the sources of fighting spirit in an individual, and youth attitudes toward military service. Also commonly studied are the levels of technical skill and education among recruits, especially during this period of increasing weapon system complexity.

POLITICAL SCIENCE PERSPECTIVE

Dr. Lebow emphasized that manpower questions can never be divorced from considerations of process. He pointed out that, in America, the raising and maintaining of military forces is at heart a political game, and the military should accept this reality from the outset. In addition, each of the various subcomponents of the manpower process are political—the bargaining among the services over force levels, reserve group lobbying to protect vested reserve interests and high end-strengths, and so forth. Lebow also asserted that, historically, force levels have been treated as a residual factor in national strategy. Congress and the Department of Defense first decide what forces the nation can plausibly afford, and then they tailor the threat to fit those forces. The whole process of determining force levels and hence, manpower policies, involves a tradeoff between the necessary and the possible.

Lebow then turned to the broader problem of maintaining U.S. public support for the commitment of military forces, large or small, to distant regions of the world. He pointed out that it is difficult for democracies like the United States to engage in large-scale military operations without substantial public support. To be accepted by the public, such operations have to be explained and justified in detail. He also argued that U.S. policymakers believe that U.S. foreign policy requires the symbolic example of a demonstration of force to be effective. In Lebow's view, resolve is a function of interests rather than of bargaining skills or military might.

ECONOMICS/ECONOMETRICS PERSPECTIVE

At the outset, Dr. Cooper noted the preponderance of economists and economically oriented analysts in the military manpower field. He conceded that this perspective has perhaps been too dominant in recent years, resulting in a struggle between the established economists and the proponents of other disciplines. At the heart of this struggle is the former group's tendency to analyze manpower issues in terms of easily quantifiable factors, an approach opposed by the latter.

Cooper argued that it is not a bad thing to base manpower analysis on hard, empirical evidence, although he admitted that this approach may lead analysts to ignore important unquantifiable issues altogether. As an example he cited the excessive emphasis of the Gates Commission on quantifiable factors, such as military pay and benefits, to the exclusion of other considerations.

In conclusion, Cooper identified two issues that can be expected to dominate economic analyses of manpower problems in the future. The first concerns the costs, in terms of pay and benefits, of various alternative manpower programs (the all-volunteer force, a fully drafted force, or a partially drafted force). The second and more important challenge is the need for noneconomists to participate fully in manpower analyses. Given the economist's inherent discomfort with the unquantifiable, such participation is essential to the development of a balanced and realistic consensus on manpower questions.

PRACTITIONER'S PERSPECTIVE

General Gorman conceded that, with regard to the manpower-strategy linkage, the situation in the Pentagon is poor but not hopeless. He noted that in organizational and conceptual terms, the Joint Chiefs of Staff have not been very active in manpower policy. One key flaw in the Defense Department's approach to strategy has been its inability to consider manpower needs and concerns on a joint basis. In effect, the manpower policies of the various commanders in chief have been determined by the individual services and the

Office of the Secretary of Defense. However, some progress in this area is being made, Gorman noted, mainly through the activities of the Defense Resources Board.

On a more general level, Gorman spoke of the need for a realistic and comprehensive assessment of the threat and of the size and character of the military forces needed to meet that threat (regardless of whether that force is actually created). Such an assessment would give U.S. manpower planners an idea of what various manpower strategies might cost and a better sense of the inadequacies of our present strategy and force structure. Gorman also argued that a strategy of deterrence implies not only a large active force structure, but also a force structure of high quality. In his view, a few good men under arms are preferable to a huge mob of inexperienced or ill-trained soldiers. The requirement for quality soldiers is especially important in Third World contingencies, which promise to be more frequent in the years ahead. Gorman concluded that military manpower policies need to be fashioned so as to support a strategy of conflict avoidance rather than conflict engagement. He suggested that although the Defense Department is improving its efforts to bring manpower considerations and strategic goals into some sort of harmony, it still has a long way to go.

DISCUSSION

Much of the ensuing discussion was dominated by a debate over the methodological problems involved in analyzing manpower and strategy together. One questioner argued that the real problem in this area is not the lack of input from noneconomists but the profusion of different views and approaches on manpower questions. He claimed a need for these varying inputs to be incorporated into a synthesis that can be presented to the operators. One panelist responded that this was what the panelists had been asked to do in their presentations—to speak only to their particular disciplines. He added that specialists are and must be aware of the limits of their approach and appreciative of the contributions of other disciplines. A second participant argued that military manpower planners themselves must bear the responsibility for assimilating and synthesizing the profusion of advice and research they receive from civilian manpower experts. A third participant claimed that the civilian political leadership must play a role here as well. Finally, it was asserted that the analysts, whoever they are, must identify the constituency they are trying to reach and structure their work accordingly.

A second questioner was concerned with future military manpower policies. Claiming that the health of the all-volunteer force is inversely proportional to the health of the civilian economy, he argued that the nation needs a viable alternative to the present manpower system.

Returning to the issue of methodology and the need for a synthesis on manpower-strategy issues, a third participant pointed out that many academicians want their work accepted *in toto* by policymakers. In reality, however, the latter select those arguments and findings that best meet current policy needs and ignore the rest. Continuing this theme, another participant urged academicians to accept the limits of their education and to break out of the arbitrarily devised academic straightjackets imposed on them. He also underlined the need for academicians of all disciplines to talk to each other continually. These assertions drew a partial rebuttal from the panel. Claiming that most academicians are in fact enthusiastic communicators with one another, they argued that the real intellectual gulf lay between the areas of manpower and strategy. It was precisely this gap that the present conference addressed. Finally, it was noted that the Department of Defense, particularly the Joint Chiefs of Staff, is inherently ill equipped to link manpower and strategy.

Another participant argued that no real growth in the defense budget for fiscal 1986, and the prospect of very limited growth for succeeding years, constitutes a significant threat to the viability of the all-volunteer force. If U.S. national security is to be preserved, then new manning initiatives must be explored seriously. This problem is complicated by the fact that America's European allies are facing the twin prospects of little or no real growth in their defense budgets and a steep decline in manpower pools.

The next questioner noted that the participants at this conference were engaged in a political process as well as an intellectual one, and he asked the panel to outline a strategy for harmonizing manpower and strategic considerations in practice. In response, it was suggested that linking this cause to the larger movement for reform of the Defense Department might bring results, although no specific recommendations for effecting such a union were offered.

The final question posed to the panelists concerned the historical studies supervised by Colonel Dupuy. One participant was interested in the reception these studies had received and wanted to know specifically how qualitative and quantitative factors were integrated in this research. Dupuy responded that the studies had received a mixed reception from the academic, operations research, and military communities. The methodology used was basically a top-down approach, in which one isolates the various inputs into a battle (numbers of troops, guns, ammunition, etc.) and then compares, in quantifiable terms, what should have happened to what actually occurred. In this way, it is possible to develop a relationship between inputs and outputs on the battlefield, using historical events as the data base.

This roundtable discussion provided conclusive evidence that, even though some sizable intellectual barriers remain to be overcome, there is a growing recognition throughout the analytical community that we have not equipped ourselves well to deal with the strategic dimension of military manpower. Mere recognition of this shortcoming is itself a noteworthy achievement. The challenge ahead will be to effect some much needed change in our analytical approaches and methods.

11 CROSS-NATIONAL ASSESSMENTS OF THE MANPOWER-STRATEGY INTERFACE

Karen A. McPherson

The ways that other countries address the manpower-strategy relationship can contribute to a generalizable, rather than U.S.-specific, understanding of this relationship. A roundtable entitled "Cross-National Assessments of the Manpower-Strategy Interface" was conducted to place the American approach to mapower and strategy in a broader comparative context. Dr. Samuel F. Wells, Jr., Associate Director of the Woodrow Wilson International Center for Scholars and Director of its International Security Studies Program, chaired the roundtable. Participants in the roundtable were the following:

- Dr. Michael J. Deane, Director of Research in the Russian Studies Center of Booz, Allen & Hamilton, Inc., and Adjunct Professor at the American University, who provided the Soviet perspective
- Dr. Michel L. Martin, Director of the Centre William I. Thomas and Professor at the Institute d'Etudes Politiques at the Social Science University of Toulouse, who provided the French perspective
- Dr. Ralf Zoll, Professor and occupant of the Chair for Applied Sociology at Philipps-University of Marburg, who provided the German perspective
- Dr. Gavin Kennedy, a member of the faculty of Heriot-Watt University in Edinburgh, Scotland, who, based on his previous

studies of Third World military capabilities, provided a Third World perspective

MANPOWER POLICY OVERVIEW

Manpower policy refers generally to the way a society recruits, utilizes, and maintains manpower in its military institutions. To establish the context for the discussion to follow, a brief overview of the substance of military manpower policy for each nation represented at the roundtable is in order.

Soviet Union

The Soviet Union uses conscription to man its armed forces; service in the army and air force is for two years, in the navy and border guards from two to three years. Soviet law requires universal male military service; this has established a military manpower base in which able-bodied male citizens between the ages of eighteen and fifty are either on active duty or subject to reserve service.[1] The total strength of the armed forces is slightly higher than 3.5 million men under arms. Total reserves are estimated at 25 million, of which some 5 million have served in the last five years.[2] About 1.8 million recruits are conscripted annually into the armed forces.[3]

France

France employs national conscription to man its armed forces; the period of service is twelve months (eighteen months for overseas service). Approximately one-half million French citizens are under arms, and about half of these are conscripts. There are 13,000 women in the French military. Army reserves total 305,000; navy, about 20,000; and air force, about 58,000.[4]

Germany

The German armed forces are manned through a combination of volunteers and conscripts, with about one-half million under arms (of which about one-half are conscripts). This manpower strength is determined by law and was initially set forth under the terms of the Western European Union (WEU) treaty, which addressed issues of German rearmament in 1955. Mobilization strength is 1.25 million. The term of military service is fifteen months.[5] Through the 1970s and into the 1980s, conscripts have constituted approximately 45 percent of total active manpower.[6]

Third World

Because the distinctions among Third World nations are at least as significant as their similarities, there is no real entity called "Third World military manpower policy" to address. However, the dimensions of the issue can be identified. The *Military Balance* shows that around two-thirds of the Third World nations for which information about manpower policy is provided employ some form of conscription or selective service, for terms ranging from six months to three years.[7]

DETERMINANTS OF MILITARY MANPOWER POLICY

Strategy and manpower policy are both dependent variables—but it is not at all clear that they are dependent on the same independent variables. Strategy appears to be more closely related to a nation's perceptions of threats to its interests, or the functional imperatives it must address. Military manpower policy seems more closely related to societal or domestic constraints, including political culture, available population, and a nation's economic condition.[8]

For example, it is clear that the United States moved from the draft to an all-volunteer force in 1972 primarily because of domestic constraints rather than functional imperatives. Other chapters in this book have made the case that, for the United States, domestic constraints determine manpower policy in virtual isolation from consideration of the functional imperatives that influence strategy. The purpose of this roundtable is to examine whether the same strategy-manpower divergence occurs in other nations.

Functional Imperative: Threat Perception

Stories are told in the United States of the lines that formed down streets and around corners at recruiting offices the day after the Japanese attack on Pearl Harbor. It is undeniable that, to some extent, people rally 'round the flag when they perceive their nation as threatened. That this action influences military manpower *policy*, as opposed to military manpower *numbers*, is not, however, readily evident.

Soviet Union. Soviet strategy can best be capsulized as victory-oriented strategic deterrence. This strategy is consistent with Soviet threat perceptions. Both by doctrine and by experience, the Soviets see enemies everywhere. Leninist doctrine warns to beware of enemies; their sobering historical experience—the allied intervention during the Soviet Civil War, the Nazi invasion of 1941, the postwar period of "capitalist encirclement," and the emergence of threats from the East in the form of Chinese nuclear power and anti-Soviet feelings—reinforces these beliefs.[9] So ingrained is the habit of portraying the West as a constant and significant threat to the Soviet Union that even the years of detente saw only an attenuation, not the disappearance, of confrontational rhetoric. The primary source of the perceived threat is the United States, and it is toward this adversary that Soviet strategy is directed. The Soviet Union pursues a strategy of deterrence, but its approach focuses on deterring war by developing its ability to win should war come about. The Soviets assume that a future war will be decisive, coalitional, global and uncompromising. They believe that war will begin with massive and surprise nuclear strikes, or, alternatively, that it will begin and remain conventional for a short time. They clearly do not expect it to remain conventional throughout.

These assumptions about the nature of a future war bear little relationship to the kind of military institutions the Soviets maintain. Their recognition of the possibility of a conventional phase of the war leads them to believe that wars may well be conducted by mass armies for a time but that *nuclear* forces will determine the outcome of the war. They also realize, given the changing political culture of the Soviet Union and the economic and political incapability of the system to support massive armies over an extended period in peacetime, that their manpower policy is dependent on the ability to mobilize rapidly in the event of war.

There is a second element to Soviet strategy that has a bearing on manpower. The Soviet Union has successfully employed its forces to promote its interests and expand its frontiers. Most recently, Afghanistan has been a target for Soviet expansionism. The manpower-intensive nature of this kind of war calls for the maintenance of forces not justified by Soviet nuclear warfighting strategy.

France. French strategy can be defined as *sanctuarisation elarge* (enlarged sanctuary). Despite De Gaulle's perception of threat *a tous azimuts*—from all directions—it is clear that the only real threat to France is from the Soviet Union (although there is some evidence that the threat is more political than military).

Nuclear forces are not manpower intensive, and until recently the French military focused almost exclusively on its role in deterring an attack on French soil with its tactical nuclear forces and its strategic *force de frappe.* A reliance on nuclear weapons, being technology intensive rather than manpower intensive, would suggest a manpower policy that emphasizes a small cadre of elite professional forces rather than a mass citizen army. However, the political culture of France, with its revolutionary heritage, continues to insist on the maintenance of mass armies even when the strategy to be implemented does not call for mass armies.

In recent years, the French have come to take a greater interest in the conventional defense of Europe, even to the point of conducting joint exercises with Germany. The French talk about an "extended sanctuary" concept of French defense—a defense of both French and German territory in the face of a thrust from the East. As France has begun to consider using its military forces for such conventional military purposes, the maintenance of mass citizen armies has come to make more sense. It should not be assumed, however, that the enhancement of French conventional military capabilities was spurred in any sense by the demands of military manpower; rather, it is a case of manpower policy staying essentially the same while policy and doctrine have caught up with it.

The French Force d'Action Rapide (FAR), established in 1983, is further evidence of a move to a more manpower-intensive form of warfare; it is too soon, however, to tell if the effect of the creation of five divisions of light, flexible, and mobile forces will have a long-term effect on French manpower policy.

Germany. German strategy is primarily to seek security through membership in NATO. This derives from a broad consensus that the Soviet Union and its allies are the major direct threat to Germany. Since 1954 the Federal Republic of Germany has focused on the conventional defense of its borders and its territory. Military manpower policy has been consistent with this focus, encouraging extended military experience in as much of the population as feasible, given the statutory limit on the size of the active force.

There is not a significant fear of direct Soviet invasion; however, Germans are aware that in the event of conflict Germany will be the battlefield. The more menacing threat is political—that the Soviet Union will attempt to use political means to provoke a schism in NATO and possibly isolate Germany from the alliance. In particular, Germans are concerned about Soviet efforts to weaken the link between Washington and Bonn. During the 1970s, the perceived success of detente raised German fears that the fate of Europe would be "arranged" by the superpowers. Brandt's policy of *Ostpolitik* was designed, at least in some measure, as a counter to such an unwelcome superpower agreement; *Ostpolitik* took on added significance as a way of reducing the threat from the East.

Third World. The security concerns in the Third World arise from three main sources: the superpower conflict, regional animosities, and domestic disruptions. Thus, to speak of the threat to the security of a Third World nation is to speak of a variety of concerns that vary dramatically from one nation to another.

Third World armies do not focus on the East/West superpower conflict. They simply do not have any role to play in that confrontation. Thus, even though an aligned Third World nation might feel threatened to some extent by the military might of the other superpower, this perception does not have an impact on military power or doctrine in that nation.

However, both the intra-regional animosities and the domestic problems faced by Third World nations have influenced military doctrine and policy, including manpower policy. Third World armies are organized primarily around issues that have little to do with warfighting per se. These armies play a major role in maintaining the internal security of the state, and personnel are recruited on that basis. They also play a role in nation-building, particularly in the postcolonial period; the educational and acculturation roles the army plays in these situations detracts from its ability to train for and to fight a war. Armies in the Third World also exist for symbolic reasons of state — to exhibit the power of the state and to accord it some legitimacy. Armies that are clothed, trained, and equipped for display purposes may be totally unprepared for warfighting. Finally, in the Third World the army is often part of a state cadre that actually participates in the political process of the nation and fills leadership positions. Filling this political role demands different recruitment criteria than those demanded for warfighting.

Given the fact that Third World militaries hold a place in society very different from that held by militaries in the Developed World, it is hard to compare the threat perceptions behind the policies, including manpower policy, of the various armies. It is accurate to say that the threats to which Third World militaries are best prepared to respond are political rather than military in nature, which suggests that manpower policy is driven more by political than by military threat perceptions.

Domestic Constraint: Political Culture

Broadly speaking, *political culture* refers to the attitudes a population holds toward its political system. The political culture of a society consists of the empirical beliefs about expressive political symbols and values and other orientations of the members of the society toward political objects. It is the product of the collective history of the political system and the life histories of the individuals who make up the system.[10] Feelings of identification, legitimacy, participation, institutionalization, and national integration are all components of a nation's political culture and can have a dramatic effect on military manpower policy. Political culture is discussed below with reference to each nation represented in the roundtable.

Soviet Union. In the Soviet Union the acceptance of the military as a legitimate arm of civic education and political indoctrination leads to maintenance of a larger military establishment than might otherwise be deemed necessary. The deeply held Russian patriotism toward the motherland—although not necessarily toward the Soviet State—has made military service in the name of the motherland a respected and honored profession. However, as memories of the revolution and of the Great Patriotic War fade, a political culture more accurately described as apathetic and pacifist has begun to emerge. There are several causes for this:

1. The society is gradually becoming more affluent. As avenues for upward mobility proliferate, the importance of the military's role in this respect diminishes. In addition, as consumerism spreads through the Soviet society, interest in the material rewards of one's career choice increases; the military does not provide such rewards and thus is less attractive.
2. Soviet peace propaganda has been almost too effective. Depictions of the horrors of nuclear war and the affirmation that war can be avoided have perhaps dulled the edge of the Soviet people's sensitivity to the need for continuous vigilance; interest in joining the military has been lessened commensurately.
3. The gradual revival of religion in the Soviet Union—certainly not sufficient yet to be called a trend, but still significant—has led to demands for religious exclusion from military service. This has also marginally diminished interest in military service.

France. The revolutionary heritage of France leads to a strongly held attachment to the concept of a mass army—a nation in arms—as opposed to an elite force. France's attitudes toward the military are strongly influenced by the Napoleonic period, by Clausewitz and Jomini, and by the two World Wars, all of which led to the development and maintenance of mass armies and to a rabid attachment to the concept of civilian control of the military. The ambiguous perception of the armed forces as both the spirit of the nation and a potential threat further complicates the French view of the value of military service.[11] There appears to be little interest in France in abandoning the mass army manpower concept, even in light of a strategy that for two decades has emphasized technology-intensive, rather than manpower-intensive, nuclear forces.

Germany. German political culture is skewed by the post-World War II occupation of the German states by representatives of the victorious Allied nations—the United States, France, Great Britain, and the Soviet Union. The legacy of Prussian military tradition, the unique role of the military in the downfall of the Weimar Republic, and the issue of military responsibility for the crimes of Nazism further cloud perceptions.[12] One of the terms of the surrender of Germany in 1945 was its complete demilitarization. By the mid-1950s, domestic and international political realities had led to the rearmament of both German states and to their inclusion in their respective European alliances. It was always clear, however, that Germany would depend ultimately on the deterrent threat provided by the United States. The Bundeswehr, as reformulated in 1956, was established to be intentionally discontinuous with the past. The postwar leaders of the Federal Republic of Germany were insistent on strict civilian control over the military, or what they called *Innere Fuhrung*, as the core organizing principle for the German armed forces. This policy rejected both the Prussian model of an aristocratic, autonomous military establishment and the Nazi model of a politicized, submissive military.[13] Adenauer spoke for all the major parties in the Federal Republic when he declared in a parliamentary debate in 1954:

> In Germany ... the army will be subordinate to the law that will be passed by the Bundestag. All those who exercise political responsibility in Germany will jointly supervise the implementation of that law. In the present era the army no longer occupies the central position that it had in the old form of society and government. The officer corps is no exclusion association that pursues its own political ambitions and in critical historical moments holds the fate of the nation in its hands. ... The army [fulfills] important functions in democratic society, but ... does not rule it.[14]

Third World. Any discussion of the Third World must be carried on in full cognizance of the fact that, as Dr. Kennedy said during his presentation at the roundtable, "generalizations are hostage to the exceptions." With this caveat in mind, however, there are a few aspects of political culture in the Third World that have implications for military manpower. In many Third World nations, for example, it is not only accepted but desirable for the military to play a significant political role. The military is accorded a degree of legitimacy in this role that would not be forthcoming in the other nations discussed here. In addition, because of the colonial background of many Third World nations, the military is often the only institution that has a nationwide base of support; remnants of ethnic conflicts that predated and coexisted with the colonial period, as well as the predilection of colonial powers to draw lines on a map and call the result a nation, led to situations where the most deeply held feelings of identity had little to do with a national identity. Only in the military was national identity promoted and, thus, success in the nation made possible. This significant difference in the roles accorded the military makes its attractiveness as a profession quite different from what it is in other nations and makes comparisons difficult. The appalling inefficiency of Third World armies—an inefficiency driven by the diversity of roles they are called upon to play—suggests that perceptions of purely military capabilities have little to do with recruitment, training, retention, or use of soldiers.

Domestic Constraint: Population

In the end, it is the numbers of people who meet the criteria set for service in the military that will determine how manpower policy is implemented—how a nation attracts and keeps the right quantity and quality of recruits. In a nation rich with manpower having the relevant characteristics (characteristics determined, as we have seen, largely by a nation's political culture), manpower policy consists of little more than waiting for people to walk in the door. In other nations, where conflicting opportunities make military service only one among many choices for large numbers of people with the relevant charactcristics, military manpower policy must be cleverly constructed to reach those people who can best meet the needs of the institution.

Soviet Union. Evidence suggests that the fortunes of Soviet military manpower availability have waxed and waned over the years. From 1970 through 1978, there was a gradual increase in the number of draft age youth; this was reflected in a slow, but steady, growth in military manpower. In 1978-79, there was a decline in manpower strength; in 1979-82, an upswing; in 1983, another cycle of decline and recovery. Such swings in the fortune of military manpower over this short a term could not have been driven by population ebbs and flows that reflect an overall annual population increase of .9 percent.[15] Projections show that the natural growth in the labor force will drop from around 2 million people to around 400,000 by the end of the 1980s, with most of the growth coming among the less skilled and less mobile Turkic population in Central Asia and Transcaucasus.[16]

France. The population of France has grown at an annual rate of .5 percent over the last five years. Of the current (1983 figures) population, 14 million males fall into the fifteen to forty-nine year-old group; of these, 12 million (85 percent) are eligible for military service. A total of 430,000 young men reach age eighteen annually.[17] For the near future at least, France appears to have an adequate supply of young men to serve its manpower needs. Given the short term of service (twelve months), any moderate manpower shortfalls could be addressed by lengthening the term of service without running the risk of establishing the kind of professional elite corps the French political culture would not accept.

Germany. The Federal Republic of Germany is facing a crisis in its military manpower policy. Given a declining birth rate, the Bundeswehr will lack 50,000 to 100,000 conscripts in the next decade. The German population is expected to decline overall over that same period. Present projections suggest that there will be a manpower deficit through the 1990s.[18] The immediate impact will be felt in the number of available conscripts, but parallel effects can be expected in the ranks of both the short-term volunteers and regular forces. Possible remedial actions—a longer conscript term, greater incentives for volunteers, a change in the conscript-enlistee ratio—give rise to some basic questions about the role of the armed forces in German society. These questions will ultimately depend on German political culture for their resolution.

Third World. As opposed to the rest of the nations discussed in this roundtable, Third World nations do not face a problem with regard to population. In the Third World there is a more than adequate supply of young people in the age cohort most efficiently used in the military. Military manpower policy must deal not with how to get people into the military, which is after all one of the main avenues of social mobility in these societies, but with how to use them efficiently in pursuit of the modernization and development objectives of the state.

Domestic Constraint: Economic Strength

The military is frequently seen as an employer of last resort. A nation with a robust, growing economy can provide jobs for its people, and thus the military must compete in the marketplace for volunteers or resort to conscription.

Soviet Union. The Soviet Union employs conscription to man its mass army, but the attractiveness of the military as a career is affected by the opportunities available throughout the society for social mobility and economic security. There is some evidence that, as the Soviet economy grows and the society becomes more affluent, the alternatives to military service will become more attractive to people who might otherwise have made a career out of military service. Recent trends suggest that, although the society is growing more affluent, the rate of growth in GNP is slowing.[19]

France. From 1959 to 1973, the French economy grew at an annual rate of 5.5 percent. Since the 1974 energy crisis, the rate of growth has averaged 2.4 percent. In the early 1980s, deficit and inflation pressures led to currency devaluation, which was backed up by spending cuts and increased taxes that further dampened economic recovery. In the mid-1980s, unemployment was around 8.8 percent.[20] This combination of circumstances presents the classic case of a situation in which an employer of last resort—possibly the military—would find its recruitment and retention problems easier to address.

Germany. The German military relies on conscription to man its armies. But retention of the manpower thus conscripted is dependent on the strength of the domestic economy. There is evidence that, as German unemployment increases, willingness to make a career out of military service will also increase. The German economic miracle appears to have lost some of its luster. Postwar rapid growth rates in GNP slowed with the 1974 oil shocks, and GNP actually fell by 3 percent in 1975. Growth has been erratic since that time. A 1985 unemployment rate of 9.2 percent reflects the general malaise of the German economy during the mid-1980s.[21]

Third World. The extreme economic inequalities that characterize the Third World give credence to the supposition that the military is an avenue to prosperity. Third World nations may or may not rely on conscription; but it is clear that the economic opportunities afforded people who serve in the military, opportunities that might be available either during or subsequent to their terms of service, make the military an economically attractive career option. Some of the more developed Third World nations have reached the point where alternative avenues to prosperity have gradually opened— career civil service, business, finance, and the like. But it is nonetheless true that in the absence of evidence of a higher level of economic development, service in the military is driven to some extent by the promise of a place to sleep, good clothing, an adequate diet, and training in the skills of an industrialized world.

MANPOWER AND STRATEGY

What then can be said about the relationship of military strategy to military manpower policy? There is no evidence that manpower needs or demands have any appreciable impact on strategy; neither is there any compelling evidence that strategy has a significant effect on manpower policy. Of all the factors discussed, a nation's political culture has the greatest impact on manpower policy. To whatever extent a nation's strategy is at odds with its political culture, it will be at odds also with its manpower policy. Because strategy is more closely linked with functional imperatives, such as threat perceptions, than with societal constraints, such as political culture, there will be a strategy-manpower mismatch. One need only look at the Soviet adherence to a policy of universal military service, which clearly serves the educational and indoctrinational needs of the state more than the needs of a military establishment that believes that any future war involving the major powers will go nuclear at an early stage; or at the French insistence on maintaining a nation at arms during two decades when its primary strategic options were nuclear and thus not manpower intensive; or at the continued German adherence to a small active armed force (limited by law to 500,000 people) when its overwhelming focus is conventional war, by nature manpower intensive, and when its economy shows the slowed growth rates and increased unemployment that might be expected to lead to greater quantity and quality of military manpower.

The cross-national perspective provided by this roundtable lends credence to the conclusion, drawn from examination of the American military manpower experience, that the strategy–manpower mismatch is, if not unavoidable, at least observable in a broader comparative context.

NOTES

1. "Sustainability, Readiness, and Mobility," *Defense Update* 49 (1984): 48.
2. *The Military Balance, 1984-85* (London: International Institute of Strategic Studies, 1985), p. 21.
3. Edward L. Warner, "The Defense Policy of the Soviet Union," in *The Defense Policies of Nations*, ed. Douglas J. Murray and Paul R. Viotti (Baltimore: The Johns Hopkins University Press, 1982), p. 97.
4. *The Military Balance, 1984-85*, p. 46.
5. Ibid., p. 49.
6. Catherine McArdle Kelleher, "The Defense Policy of the Federal Republic of Germany," in Murray and Viotti, *Defense Policies of Nations*, p. 290.
7. *The Military Balance, 1984-85*.
8. Samuel Huntington, in his 1957 book *The Soldier and the State* (New York: Vintage Books) provides an extended discussion of the interplay of functional imperatives and societal or domestic constraints in the determination of a nation's civil–military relations, of which military manpower policy is one dimension.
9. Benjamin S. Lambeth, "The Sources of Soviet Military Doctrine," in *Comparative Defense Policy*, ed. Frank B. Horton III, Anthony C. Rogerson, and Edward L. Warner III (Baltimore: The Johns Hopkins University Press, 1974), p. 203.
10. Samuel P. Huntington and Jorge I. Dominguez, "Political Development," in *Macropolitical Theory*, ed. Fred I. Greenstein and Nelson W. Polsby (Reading, Mass.: Addison-Wesley Publishing Company, 1975), p. 15.
11. Alan Ned Sabrosky, "The Defense Policy of France," in Murray and Viotti, *Defense Policies of Nations*, p. 248.
12. Kelleher, "The Defense Policy of the Federal Republic of Germany," p. 288.
13. Ibid., p. 289.
14. M. Donald Hancock, "The Bundeswehr and the National People's Army, in Horton, Rogerson, and Warner, *Comparative Defense Policy*, p. 67.

15. *Countries of the World and Their Leaders Yearbook 1985* (Detroit: Gale Research Company, 1984), p. 1176.
16. Ibid., p. 1183.
17. *The World Factbook 1985* (Washington, D.C.; Central Intelligence Agency, 1985), p. 78.
18. Kelleher, "The Defense Policy of the Federal Republic of Germany," p. 290.
19. *Countries of the World and Their Leaders Yearbook*, p. 1182.
20. Ibid., p. 506.
21. *The World Factbook 1985*, p. 85.

VII SUMMARY AND CONCLUSIONS

12 THE STRATEGY-MANPOWER INTERFACE
Retrospect and Prospect

Alan Ned Sabrosky
William J. Taylor, Jr.

Three principal themes dominated the presentations and ensuing discussions of issues raised at this conference. One was that the strategy-manpower interface was of considerable theoretical importance and would likely become increasingly significant in the years ahead. The second—standing in rather striking contradistinction to its predecessor—was that there existed in practice a profound mismatch between manpower policy and strategic planning, a mismatch that will be extremely difficult to correct. And the third, in part a derivative of the second theme, was that the United States needs to do better in preparing for the effective conduct of future low-intensity conflicts. This chapter assesses these three themes and then reflects upon their implications for U.S. planners.

THE STRATEGY-MANPOWER INTERFACE

Demographics in general have always played a major role in the definition of national power. Relatively small states sometimes discomfited more populous adversaries; Sparta, Prussia, and Israel come immediately to mind in different eras of history. But the logic upon which Voltaire's avowed supremacy of "big battalions" rested was more commonly that which "associated power with population," a point driven home more recently by John Keegan.[1] Technology has changed many things since Voltaire's time, of course. Certainly, the events attending World War I ought to have highlighted the limitations of "manpower alone" in modern warfare. Yet there remains considerable support for the proposition that population is a necessary determinant of power at the *national* level, if not sufficient to account for purely *military* prowess.[2]

The fact that population is an essential element of national power logically suggests that it is also a concern of *strategy*, which is, among other things, concerned with the manipulation of power. This was inherent in Gregory Foster's assertion that "power is the quintessence of strategy." Irving Horowitz carried Foster's point a step further, arguing that "the human resource factor, far from being peripheral, is central to strategic-military concerns"—especially with regard to the Soviet-American military balance in an era of nuclear parity.

Paralleling the general consensus that emerged on the importance of the strategy-manpower interface was an equally general acknowledgment of its complexity. Not surprisingly, definitions of that interface varied considerably. For Mady Segal, there are four "levels" in the equation: (1) *societal*, raising the issue of the legitimacy of the military, particularly with regard to its role as a deterrent force; (2) *organizational*, with regard to the character of a nation's military institution; (3) *group dynamics*, focusing on unit cohesion and morale; and (4) *individual*, addressing the question of who joins and fights, and why. Irving Horowitz, in contrast, defined the interface in terms of three "strategic aspects of manpower": (1) relations between, and the relative strength of, contending military establishments; (2) functions of manpower (e.g., peacekeeping); and (3) goals (e.g., maintaining global equilibrium). In each instance, Horowitz asserted, manpower-related considerations would have an impact on the outcome. Foster developed Horowitz's line of argument even further by examining the "manpower dimension of strategy," or more generally, the role played by manpower in military affairs. He considers eight issue-areas to be pertinent: (1) light versus heavy forces; (2) the "tooth-to-tail" ratio; (3) the ratio of active duty forces to reserves; (4) elitism versus a mass army; (5) voluntarism versus conscription; (6) racial/ethnic homogeneity versus heterogeneity; (7) the female content of the forces; and (8) the fact, or potential, of military unionization. The relative importance of these factors will obviously vary, but all merit attention.

THE STRATEGY-MANPOWER MISMATCH

The fact that a diverse set of issues in the manpower-strategy interface *merit* attention does not mean that they actually *receive* it. In fact, the overwhelming consensus was that the linkages between strategic planning and manpower policy are, in the case of the United States, tenuous at best. Sam C. Sarkesian went to the heart of the matter by pointing out the fundamentally different concerns of those charged with strategic planning, on the one hand, and with manpower policy on the other. His conclusion—echoed by Franklin Margiotta—was that the ensuing "strategy-manpower mismatch" had "multidimensional implications" (all of them bad) for national security.

Interestingly enough, this mismatch is not at all peculiar to the United States. The cross-national roundtable, assessing this question from the perspective of the Soviet Union, France, West Germany, and selected Third World military systems, found that "the strategy-manpower mismatch is, if not unavoidable, at least observable in a broader comparative context." The roundtable participants found no evidence that manpower needs or demands have any impact on strategy in the countries under consideration. Similarly, there was no compelling evidence that strategy had a significant impact on manpower policy in those same countries. Indeed, the very universality of the strategy-manpower mismatch among countries so different in other respects suggests that some of the criticism leveled against the American military institution may have cast in *national* terms problems that are *generic* to modern military establishments.[3]

Generic or not, the strategy-manpower mismatch has pointed consequences for the United States. General (retired) Paul Gorman underscored this factor when he noted that a key flaw in the Defense Department's approach to strategy has been its inability to consider manpower needs and concerns on a joint basis. Robert Pirie elaborated on Gorman's point, observing that "strategists view manpower experts in the same way they view logisticians—with extreme distaste."[4] The result of this antipathy toward manpower specialists is, in his opinion, readily apparent in the Wartime Manpower Planning System (WARMAPS). There, Pirie remarked, "Manpower is not an input to the strategic calculus; rather it is an output, or a residual." The net effect is a tendency for strategic planners, in Horowitz's view, to assume that manpower needs will be met, even in the mobilization-type contingencies inherent in what Sarkesian called our "traditional strategy and scenarios that may be unrealistic." The possibility that even that strategy and those scenarios may ask more than our reserve mobilization system can deliver, as William Hauser concluded, simply compounds the problem.

It would be unwarranted, however, to allow that problem to be formulated in terms of innovative manpower policymakers struggling vainly against tradition-blinded strategic planners. On the contrary, as Foster astutely remarked, "Manpower analysis, in turn, has . . . [failed] to capture the hearts and minds of strategists." Even Sarkesian, whose criticism of strategic planning was especially pointed, concurred, stating flatly that "manpower planning reflects the same gaps and problems [between scenarios and realities], paralleling strategic perceptions." Horowitz saw the failure of what he labeled "broad manpower management" in this country to have gone beyond purely military considerations, as it had "neither won the internal war against poverty nor the external war [in Vietnam] against totalitarian enemies." And it is useful to reflect upon Pirie's identification of the reasons for the peripheral role of manpower issues in strategic planning as factors contributing to that failure. Those reasons are: (1) an excessive focus on "force structures" in the planning process; (2) the fact that "dollars, not people" are the principal constraint on strategic planners; (3) the "fuzzy nature of national strategy development"; and (4) the "ambiguity and imprecision that attend the use of the word *strategy*."

LOW-INTENSITY CONFLICT

The cumulative effect of these considerations, whether oriented toward strategic planning or toward manpower policy, was considered to be profound. There was virtual unanimity that "conventional perspectives pervade strategic outlooks and in turn drive manpower policy," as Sarkesian put it. Yet that conventional orientation runs counter to the prevalence of what Eliot A. Cohen has called the "small wars" of the post-1945 world[5]—wars, as Jeffrey Record indicated, that are characterized principally by their "limited intensity and geographic scope." Furthermore, it seems all too likely that these small wars, or low-intensity conflicts (LIC) as they are more commonly known, will pose a major challenge to an American defense establishment whose force structure, doctrine, and resourcing (materiel and human alike) apparently devote undue attention to the most unlikely high-intensity conflict scenarios for planning purposes.

The emphasis on being prepared to deal principally with low-intensity conflicts in the years ahead necessarily raises certain questions. One is that the American political culture—indeed, that of most Western democracies—is not particularly attuned to the conduct of protracted limited wars. Another is that the Defense Department, absent occasional exceptions such as the army's light force initiative, is dominated strategically and structurally by the least likely but potentially most destructive high-intensity conflicts. And the third, as Martin Binkin and Richard V. L. Cooper (among others) have pointed out, is that the manpower policy that coexists with the defense establishment is driven primarily by considerations of domestic politics and cost—neither of which can commonly be considered to be principles of war or canons of strategy.

THE STRATEGY–MANPOWER INTERFACE RECONSIDERED

There is an obvious need to better link manpower policy and strategic planning and to get *both* closer to *realistic* (i.e., most likely) conflict contingencies. Yet it is also necessary to note that while strategy and manpower issues affect one another, the relationship between them is not symmetrical. Strategic planning should drive manpower policy, which, in its turn, should strive to meet the requirements of that strategy. Thus, the question to be answered by those charged with *strategic planning* is, *Against whom and under what constraints are we most likely to have to fight?* The subsequent question for those overseeing *manpower policy* is, *How can the United States obtain individuals who are collectively suitable for those specific types of conflict?*

For all practical purposes, strategic planning in this context entails assigning priorities to probable missions and specifying an appropriate force structure to execute those missions. (Other factors, such as the development of a suitable doctrine, are certainly pertinent, but they lie outside the scope of this discussion.) The first element is fairly straightforward, at least from the perspective of the conference participants: this is the need to give precedence to contingencies at the lower end of the conflict spectrum, ranging from localized conventional operations to the use of special operations forces. But there is also an obvious requirement to retain some heavy forces to reassure politically (if not strategically) America's principal allies in Western Europe and Northeast Asia. This need not, however, mean that the United States *per se* must maintain the current proportion of its active duty forces committed to NATO. William J. Taylor argued for a shift of heavy unit missions to our European allies, permitting American forces to restructure increasingly for LIC missions in the Third World where our European allies cannot or will not pick up "out of area" responsibilities. In fact, Taylor suggested that it is not absurd to think about *giving* four divisions worth of U.S. prepositioned (POMCUS) stocks now in Europe to the Federal Republic of Germany for assignment to German reserve forces. The FRG faces military manpower problems of much greater magnitude than the United States and soon will be forced to draw down active duty manpower. Thus, the same *basic* mission requirements would remain. They would simply be reordered to put LIC in first place.

The second element—providing an appropriate force structure—is strategically simple and bureaucratically contentious. The experience of the past decade is that the army and the marines have moved in opposite directions to provide each service with some capability to operate across the so-called spectrum of conflict. Thus, the traditionally "light" Marine Corps has "heavied up" considerably, whereas the more heavily mechanized army has discovered the utility of the Light Infantry Division (LID). Aside from the obvious redundancy of such efforts, the violation of the principal organizational imperatives of functional specialization and division of labor reinforces endemic interservice rivalry to the general detriment of the military's collective capacity to carry out either type of mission well.

The solution that appealed to many at the conference is neither new nor without parallel abroad. What is essentially required as a matter of national policy is a functional separation of the LIC and de facto general war (or anti-Soviet) missions between the Marine Corps and the army, paralleling the different *architectures* the two forces would need.[6] The obvious difficulty here is that there is little evidence that either service—and especially the army—would relinquish easily the opportunity to participate in any contingency, simply for the sake of the national interest. Perhaps an alternative—

Table 12-1. Architecture of Basic Force Structures.

FORCE I: LIC CONTINGENCIES	FORCE II: GENERAL WAR CONTINGENCIES
a. Light	a. Heavy
b. "Toothy"	b. Big "tail" (sustainability)
c. Active components only	c. Large, mobilizable reserve components
d. Small, elite	d. Mass force
e. All-volunteer force	e. Selective service or national service
f. Homogeneous	f. Heterogeneous
g. Few if any females	g. Large female component acceptable
h. No unionization	h. Unionization could be acceptable

See Gregory D. Foster, "Manpower as an Element of Military Power," Chapter 2 in this book.

building on some precedents abroad (e.g., the French *Force d'Action Rapide*) and the unified command concept—would be to create a new unified command functionally dedicated to LIC contingencies. This U.S. Low Intensity Conflict Command (USLICCOM) would have operational control before the event over all assets applicable to low-intensity conflict contingencies. This is far from a perfect solution, but it might well be an improvement over the prevailing state of affairs.

Indeed, applying Gregory Foster's eight issue-areas noted earlier highlights the profound differences one could expect to obtain in those two forces that might stimulate reform (see Table 12-1). The relative importance of the eight factors obviously varies, and at least one—item (d), for the LIC force—could be a double-edged weapon. It is all well and good to argue for quality, not quantity in an elite force that would reject managerialism and reaffirm its commitment to a standard of professionalism that includes the classical warrior values. In fact, such a line of argument is all but obligatory these days, as a perusal of the works of Edward Luttwak, Jeffrey Record, William Lind, and Richard Gabriel, among others, demonstrates clearly.[7] It is certainly true that a small, efficient force *can* defeat a large and inefficient force—a fact as evident in ancient warfare as in modern times. But it is also true, as John Keegan gently pointed out, that numbers *do* tell *if the large force becomes efficient*—and this development is by no means unthinkable in the Third World as the process of modernization continues.

Far more essential are the requirements that a LIC force be *light*, built solely on *active component* elements, and consist of *volunteers*. The first of these elements is not particularly contentious. Any LIC force must be able to move quickly and to have minimum lift and sustainability requirements if it is to operate in such a conflict environment with maximum success. The other two considerations may seem less obvious but in fact are also necessary. This is not because there is any inherent value in, for example, a long-service professional force in general terms. But it is evident that a critical element in the successful conduct of a low-intensity conflict is reducing the degree to which that conflict becomes politicized within the United States. A lesson of the Vietnam War that bears remembering is that conscripts cannot be used effectively in limited wars, if only because of the domestic political sensitivities their employment arouses. What held for selective service conscripts before would probably hold for national service conscripts as well. Volunteers, however, are conventionally considered professional soldiers, employable in situations where conscripts could or should not be used. Thus, relying solely on volunteers to meet military manpower requirements gives this country a wider range of military options than would obtain with some form of conscription. To be sure, using conscripts—or, more specifically, mobilizing and deploying Reserve Component units— *may* signal a greater *initial* commitment to the operation. But if success eludes the United States initially, the presence of reserves or conscripts rapidly becomes a political liability. This applies even to countries (e.g., France) whose general preference is for some form of conscription or national service. Thus, the lesson of low-intensity conflicts in the modern world is clear: *Regulars who are also volunteers buy time for an intervention to succeed.*[8]

The second general question noted above focuses on the requirements of manpower policy. There are obviously a number of variants that could be applied to the general war force, if not to the LIC force.[9] Some of those choices are a matter of cultural preference. Anglo-American societies, for example, tend to endorse all-volunteer, active duty forces supplemented by a volunteer militia. Conscription, whether selective or national, can be endured, but is hardly the system of choice. Yet in France, Israel, Sweden, and the Soviet Union, to name but a few, national service is the preferred mode. The fact that the United States now has an all-volunteer force, and seems politically predisposed to retain it, barring the certainty of institutional disaster, defines certain explicit parameters within which American manpower policy must be shaped.

Four factors seem likely to affect the survivability of the all-volunteer force and its utility as an instrument of national policy. One is the *political* character of the American system and the general political dimension of the strategy–manpower interface. This was identified by Sam Sarkesian and Richard Ned Lebow as being of particular concern, with respect both to defining strategy and to shaping manpower policy. For example, it seems politically unlikely that the United States will return to some form of conscription if it was reluctant to do so during the *very* dark days of the all-volunteer force (AVF) in the late 1970s, however much that reluctance may have been influenced by the so-called Vietnam syndrome. One may debate the relative cost, or cost effectiveness, of alternatives to the AVF. Barring institutional disaster, however, the AVF probably is here to stay for the foreseeable future.

The other three factors, as Martin Binkin suggested, can either sustain the political dimension or, in certain circumstances, work to alter it. The *demographic* dimension is known: The supply of prospective volunteers will continue to decline through the mid-1990s. This means that the volunteer military will have to attract a steadily increasing percentage of a declining pool of volunteers in order to meet its requirements. This, in turn, will exacerbate the *budgetary* dimension. The weight of opinion suggests that there is a linkage between economic recovery and a decline in *quality* enlistments that will add to the general recruiting demands on the AVF. The experience of the AVF has been that the services pay more to attract and retain quality personnel than in the pre-AVF environment. And these considerations, taken together with projected federal budget deficits and declining defense budgets under the pressures of the Gramm–Rudman–Hollings legislation of 1985, are far from encouraging. The U.S. armed services, especially the army, are likely to be faced with a traditional, periodic dilemma arising at times of declining defense budgets or imposed budget ceilings. That is, the services will be forced to make the choice between maintaining "hollow armed forces" (i.e., cutting manpower and maintaining current force structure), or reducing force structure and keeping fully manned units.

The *technological* dimension essentially reflects a predisposition to use materiel rather than manpower to accomplish certain missions.

As Foster, Pirie, and Horowitz observed, technology can be seen either as a force *multiplier* (making manpower more effective) or as a force *substitute* for manpower. The distinction is critical for both strategic planning and manpower policy. The problem is that the American preference to treat technology as a substitute for manpower (thereby reducing casualties and offsetting manpower shortages) is not likely to be operationally effective. More sophisticated technology generally requires more skillful personnel and more complicated maintenance facilities, the former in increasingly short supply into the 1990s and the latter increasingly costly. The so-called emerging technologies have a place, of course, but there is much truth to Horowitz's argument that "the armed forces cannot simply substitute sophisticated hardware to displace manpower . . . as a mechanism for holding manpower needs constant." Strategy, demography, and budgetary considerations all argue against such a course of action.

The interplay of these factors, as noted earlier, will affect the extent to which the AVF can be retained. The current condition of the AVF is certainly better than was the case a decade ago. Whether that improvement can be sustained in light of the aforementioned demographic and budgetary considerations is not entirely certain. If one assumes that major reductions in aggregate manpower requirements for the total force cannot be reduced, and the dark days of the AVF reappear, consideration may have to be given to an alternative mode of manpower procurement.

One alternative that has enjoyed some support in the Congress, and more in the military hierarchy, during the difficult days of the AVF, is a return to some form of draft. Proponents of this alternative argue that a more equitably administered selective service system could meet this country's military manpower requirements fairly, effectively, and relatively inexpensively. An explicit return to some form of conscription, however, seems unlikely at this time, although a standby draft mechanism that goes beyond mere registration can—and should—be set up. Younger Americans, who would be directly affected by a reinstitution of the draft, hardly seem enthusiastic about the idea. Congress as a whole is not likely to endorse so politically sensitive a measure if such a step can be avoided, even though it may have given its assent to the registration of young people as a first step in that direction.

A more popular alternative to a draft is what is called the national service concept. Two rather different groups endorse this particular approach. Those who believe the AVF is in the Indian summer of its existence and fear that it will become a failure look to national service as a replacement for a noble (or ignoble!) experiment that failed. But there are also those who believe that the AVF is working and will continue to work in the future, but still see national service as a potentially more equitable and efficient means of meeting defense manpower needs and instilling a sense of national spirit.

There are a number of different national service schemes, but all have certain points in common. All would register young American males (and, in some cases, females as well) on or about the age of high school graduation. Those registered could choose either military or nonmilitary options, within the limits imposed by manpower requirements at that time. Shortfalls in any national service institution would be eliminated by the involuntary assignment of newly registered personnel. Any of the registrants not needed in a given year would be placed in an on-call manpower pool for a specific period of time.

At first glance, the concept of national service is appealing. It would be reasonably equitable. All would be called and all would perform some kind of service, with compensation and post-service benefits commensurate with the service performed. It would eliminate shortfalls in the armed forces, and it might generate a greater sense of national spirit than many believe now exists. Under the employer-of-last-resort concept, it would also reduce unemployment levies. Despite its initial attractiveness, however, the concept of national service merits closer scrutiny than its advocates might wish. The case for national service essentially rests on the propositions that national service would be less costly than the AVF and less compulsory than the draft.

Neither of these propositions can withstand careful appraisal. In the first place, manpower costs for the military alone would at least equal those of the AVF, even discounting the additional expense of staffing and operating a national service bureaucracy. Post-service benefits for one-term national service people would exceed those now granted to most military volunteers and a return to the inadequate pay scales of the pre-AVF era to offset those added costs is unlikely. It would be utterly unrealistic to expect any Congress not bent on political suicide to simultaneously call up entire year-groups of American youth and significantly reduce their wages below current levels. The cost of the nonmilitary component of national service would obviously increase the total expense of the scheme. Studies done in the late 1970s indicated that a fully all-male national service program would add an estimated $15 to $25 billion to the federal budget, and a program including both men and women would double that figure.[10]

Second, as far as the military would be concerned, national service would be functionally indistinguishable from a draft. Indeed, it is specious to draw a distinction between voluntary and compulsory variants of military national service systems. If a voluntary national service scheme could succeed, there would be no need for it, as the AVF itself would be able to attract sufficient volunteers. There may well be support among the youth for some nonmilitary form of national service (e.g., within ACTION), analogous to the appeal generated by the Peace Corps in the early 1960s. But there is absolutely no evidence to support the contention that there is a corresponding interest in spending two years in the infantry.

In addition, past and potential shortfalls in military volunteers could even be expected to increase if the United States turned to some form of national service. Many now enlist for pay, occupational training, and post-service benefits. If some pay, job training skills, and benefits could be obtained by nonmilitary national service, there would be much less incentive for potential recruits to opt for the more demanding, and potentially more dangerous, *military* national service in pursuit of marginally greater benefits. This would obviously compound the military's current recruiting problems and increase the number of national service people who would be assigned involuntarily to the armed forces—that is, who would be drafted for military service.

In sum, conscription in any guise is not the way the United States should meet military manpower requirements. It is a form of indentured servitude, and as such is unlikely to provide soldiers capable of meeting the militarily demanding and politically sensitive challenges likely to confront this country in the future. Second, national service is simply another form of conscription whose somewhat greater equity would be offset by significantly greater costs. It is therefore not a viable alternative to the AVF. Thus, if shortfalls of consequence reappear in the future, and tinkering with the AVF (e.g., lowering quality standards to maintain manpower levels) is properly rejected, then manpower levels may have to be reduced in the active component and greater reliance placed on reserve component forces. Whether this can, or *should*, be done is the subject of another paper. Yet the key thing to remember is that whatever changes do occur should be the result of strategic guidance, and not merely a function of marginal bureaucratic adjustments in the face of adversity.

CONCLUSION

Clearly, the strategy-manpower interface is both complicated and in need of clarification. As Sam Sarkesian put it in his chapter, "Strategists and manpower planners need to develop a common set of strategic concepts and manpower premises, with the idea that the winner in most contemporary conflicts may not be the one who gets there first with the most, but the one who gets there quickly with the best." Therein lies the challenge to be met in the years ahead.

NOTES

1. Here, and elsewhere, positions associated with specific individuals refer to their contributions to this volume unless specific documentation to the contrary is provided.
2. Katherine Organski and A.F.K. Organski, *Population and World Power* (New York: Alfred Knopf, 1962); Ray Cline, *World Power Assessment 1977* (Boulder, Colo.: Westview, 1978).
3. See military reform critiques in E. Luttwak, *The Pentagon and the Art of War* (New York: Simon and Schuster, 1984).
4. The anomaly of this is apparent in Martin Van Creveld, *Supplying War* (London: Cambridge University Press, 1977).
5. Eliot A. Cohen, "Constraints on America's Conduct of Small Wars," *International Security* (Fall 1984): 151-81.
6. William Hauser, *America's Army in Crisis* (Baltimore: Johns Hopkins University Press, 1973). See also Chapter 8 by Jeffrey Record in this volume.
7. See, for example, Luttwak, *The Pentagon and the Art of War*; Jeffrey Record, *Revising U.S. Military Strategy* (Washington, D.C.: Pergamon-Brassey's, 1984); Richard Gabriel, *To Serve With Honor* (Westport, Conn.: Greenwood Press, 1982).

8. This issue is discussed in A.N. Sabrosky, *The Politics of Military Intervention* (in preparation).
9. See a useful overview of these issues in R.K. Fullinwider, ed., *Conscripts and Volunteers* (Totowa, N.J.: Rowman & Allanhild, 1983).
10. See Richard V.L. Cooper, "AVF vs. Draft: Where Do We Go From Here?," in *Defense Manpower Planning: Issues for the 1980s*, ed. William J. Taylor, Jr., et al. (New York: Pergamon Press, 1981), p. 98. See also Martin Binkin, *America's Volunteer Military: Progress and Prospects* (Washington, D.C.: The Brookings Institution, 1984), pp. 58-59.

INDEX

Adenauer, Konrad, 201
African regiments in French army, 45
Air Force, 146
  technology modernization in, 102
  women in, 111
AirLand Battle doctrine, 133, 134, 158-159
Algerian War, 156
Antiwar literature, 166
Ardant du Picq, Charles, 156
Armed Forces Qualification Test (AFQT), 126
Army, 146
  career/first-term ratio in, 61
  fighting ethic in soldiers in, 65-66, 73, 82-83
  future conflict environments and, 154, 157-160
  manpower procurement and, 133
  means for raising, 40-41
  New Manning System in, 159
  people relationship to, 78
  proportion of males with technical skills in, 105, 106-107
  survey of officers of, 128-129
  technology modernization in, 101
  two-tiered, 156
Austria, 46

Battlefield units, and procurement policy, 133-134
Belgium, citizen forces in, 46
Binkin, Martin, 6, 91-114, 215
Black troops
  NATO perception of, 33
  see also Racial composition of military
Blitzkrieg, 45
Boulding, Kenneth, 22, 25, 26
Braestrup, Peter, 173
Brandt, Willy, 198
Budget considerations, and manpower policy, 14

Capital, and military applications, 61
Career officers
  efforts to retain, 127-128
  first-term ratio with, 61, 112
  procurement policies and, 127-131
  reform of system for, 135-136
Carter administration
  military pay in, 93
  strategy formulation in, 54
Caste, and army membership, 41
Center for Strategic and International Studies, 3-4
Central America, 148, 151

225

INDEX

Ceremony, and military power, 27, 28–29
China, People's Republic of, 20, 179
Citizen army concept, 40, 187
  European use of, 46–47
  U.S. use of, 186–187
Civilian Conservation Corps, 164
Civilian employees
  human resources planning and, 175
  manpower procurement policies and, 111, 122
  political leadership and, 190–191
Class, and army membership, 41
Clausewitz, Karl von, 22, 39, 68–69, 76, 80, 186, 200
Cohen, Eliot A., 168–169, 214
COHORT system, 133
Combat capabilities
  factors affecting, 66–67
  perceptions of, 30
  women troops and, 123
Commune, 42, 44
Compulsory service, *see* Conscription
Computers, and manpower procurement, 101–102
Conflict spectrum, and manpower planning, 74–76
Congress
  conscription and, 113, 220
  force levels and, 188
  manpower planning and, 78–79, 83
  manpower procurement and, 110–111, 135
  military pay and benefits and, 93
  national service concept and, 221–222
Congressional Budget Office, 108
Conscription, 41
  comparison of recruit test scores in, 93, 94, 95
  compulsory, 46
  French experience with, 44–45
  German army of 1913 as example of, 42–44
  introduction of idea of, 40, 41–42
  past debates over, 92–97
  perceptions of, 31–32
  return to use of, 112–113, 134–135, 173, 220
  sharpshooter movement and, 46
  troop acceptance of, 44
  U.S.–Soviet relations and, 114

Conventional weaponry, 144
  light versus heavy forces in, 30
  power and, 17
Cooper, Richard V. L., 185, 189, 215
Costa Rica, 151
Costs
  alternatives to manpower program and, 189
  military retirement and, 130
  national service concept and, 221–222
Cross-national assessment, 193–206, 213
  determinants of military manpower policy in, 195–205
  economic strength constraints in, 204–205
  manpower policy overview in, 194–195
  political culture constraints in, 199–202
  population constraints in, 202–203
  threat perception and, 196–199
Cultural factors
  military power and, 27–29
  racial composition of military and, 33

Deane, Michael J., 193
Decentralized command authority, 157
Decisionmaking
  image and, 22–23
  manpower and, 14
Deep Strike concept, 20, 133
Defense Officer Personnel Management Act of 1980, 128
Defense Resources Board, 190
De Gaulle, Charles, 197
Deitchman, S. J., 61
Delayed Entry Program (DEP), 121
Democratic Party, 166, 174
Demographic changes
  human resources planning and, 173
  manpower procurement and, 97–99
  population projections in, 97, 98
  proportion of qualified and available males required and, 97, 99
  volunteer force and, 219
Department of Defense
  manpower requirements of, 55
  manpower strategy of, 189–190

Deterrence policy
  combat effectiveness and, 66, 67
  future conflict environments and, 152
  military retirement system and, 131
  nuclear weapons and, 145-146
Dewey, John, 165
Discipline, and manpower efficiency, 38-39
Dominican Republic intervention of 1965, 146
Draft, *see* Conscription
Dress, military, 27-29
Dupuy, Trevor N., 185, 186-187

Eccles, Henry, 15-16
Econometric perspective on manpower, 189
Economic conditions
  cross-national assessment of, 204-205
  human resources planning and, 172-173, 180-181
  manpower procurement and, 93, 96, 105-108, 121, 154, 191
Edelman, Murray, 23
Educational benefits, military, 93
Elite forces, perceptions of, 31
El Salvador, 148, 151
Enlisted Personnel Management System (EPMS), 118
Ethnic groups, *see* Racial composition of military

Falkland Islands, 147, 150, 155, 157
Federal Republic of Germany, *see* West Germany
Female troops, *see* Women troops
Fighting ethic in soldiers, 65-66, 73, 82-83
Fiore, Quentin, 25, 26, 27
Follow-on Force Attack concept, 20
Force, power differentiated from, 18
Force d'Action Rapide (FAR), 197-198, 217
Ford administration, military pay in, 93
Ford Foundation, 4, 166
Foreign policy, and human resources planning, 174-175
Foster, Gregory D., 3-9, 4, 13-34, 212, 214, 217, 220

France
  African regiments in army of, 45
  Commune in, 42, 44
  conscription in, 40, 42, 44-45
  cross-national assessment with, 194, 197-198, 200, 203, 204, 213
  economic strength constraints in, 204
  German confrontation with, 44-45
  importance of troop strength in, 37
  manpower mobilization in World War I and, 47
  manpower policy overview in, 194, 205
  Napoleon in, 39
  political culture constraints in, 200
  population constraints in, 203
  threat perception in, 197-198
  two-tiered army in, 156
  volunteerism in, 47
Frederick the Great, 38, 39
Future conflict environments, 143-160
  aspects of, 150-152
  decentralized command authority in, 157
  historical perspective in, 144-147
  implications for policy of, 152-157
  possible settings in, 147-150
  prospects for changes with, 157-160
  quality of manpower and, 153-154
  small-unit cohesion in, 155-156
  specialized training and warrior values in, 154-155
  spectrum of conflict in, 144
  strategic planning and, 215-216
  Third World and, 147-148

Gabriel, Richard, 217
Garrett, James M., 178
Gates, John M., 169
Gates Commission, 189
Georgetown University Center for Strategic and International Studies, 3-4
Germany
  conscription and army of, 42-44
  French confrontation with, 44-45
  manpower mobilization in, 47-48
  World War II and, 186
  see also West Germany
GI Bill, 93

Ginzberg, Eli, 164
Goffman, Erving, 23-24
Gorman, Paul F., 186, 189-190, 192, 213
Grace Commission, 119, 129, 130
Great Britain, conscription in, 46, 47
Grenada invasion, 146, 148, 150, 153
Guerrilla war, 148, 167, 170

Hart, Gary, 113, 174
Hastings, Max, 65
Hauser, William L., 6, 117-136
Henderson, William Darryl, 73
Hertz, Heinrich, 24
Hispanic troops
  human resources planning and, 175
  NATO perception of, 33
  see also Racial composition of military
Holland, citizen forces in, 46
Hollings, Ernest F., 113
Horowitz, Irving Louis, 7, 163-181, 212, 213-214, 220
Howard, Michael, 15, 68-69, 76, 80
Human resources, 163-181
  alliances and, 179
  antiwar literature and, 167
  components of, 163-164
  conscription renewal and, 173
  economic philosophy of government and, 180-181
  economic pressures and, 172-173
  foreign policy and, 174-175
  future characteristics of, 180
  historical record on, 164-167
  material resources versus, 170-176
  national planning and, 172
  pay and, 175-176
  racial composition of force and, 175
  technological parity and, 177-178
  U.S.-Soviet balance and, 176-179
  Vietnam experience and, 167-170
  volunteer armed forces and, 166-167, 173-174
  see also Manpower procurement

Image
  control move and, 23-24
  decision in response to, 22-23
  Game of Machismo concept in, 26-27
  target group identification of, 26
  women in combat and, 32

Impression management, 23
Individual Ready Reserve (IRR), 62, 132
Iran, 148
  hostage incident in, 131, 144, 146
  possible Soviet invasion of, 149-150
Iraq, 152
Israel, 32

Janowitz, Morris, 66, 73
Jervis, Robert, 22
Joint Chiefs of Staff, 174
  decentralized command authority and, 157
  manpower policy and, 189-190
  military strategy and, 69
  readiness exercises and, 132
Jomini, Henri, 200
Junior officers, procurement of, 126-127

Kahn, Herman, 177
Kampuchea, 148
Kaplan, Abraham, 16
Keegan, John, 4, 5, 37-49, 186, 211, 217
Kennedy, Gavin, 193-194, 201
Kipling, Rudyard, 46
Knorr, Klaus, 19
Komer, R.W., 60
Korb, Lawrence J., 174
Korea, 152
Korean War, 145, 166, 186

Labor market conditions, and manpower procurement, 105-108
Lacy, James, 62
Larteguy, Jean, 28
Lasswell, Harold, 16
Lebanon, Marine Corps forces in, 131, 144, 146, 148, 150
Lebow, Richard Ned, 185, 188, 219
Leninist doctrine, 196
Lewy, Guenter, 169
Libya, 148
Liddell Hart, B.H., 18
Lilienthal, David, 165
Limited wars
  categories of conflicts and, 144
  strategy planning for, 69
Lind, William, 217
Lippmann, Walter, 165-166
List, Friedrich, 43

Low-intensity conflict (LIC), 214–215
  manpower policy and, 218–219
  unified command in, 217
  volunteers in, 218
Luttwak, Edward N., 14, 176–179, 217

Maintenance personnel, and technology, 103–104
Male bonding, 123
Managerial elite, and procurement policy, 131–132
Mangin, Charles, 45
Manpower
  characteristics of, 14
  contemporary approaches to, 51–87
  cross-national assessment of, 193–206
  discipline and efficiency in, 38–39
  discriminative principle in, 48
  doctrine and importance of, 21–22
  economics perspective on, 189
  expansion beyond national borders of sources of, 45–46
  historical perspective on, 186–187, 192
  historic overview of approaches to, 37–49
  human element in analysis of, 13–14
  importance of numbers in, 37
  low-intensity conflict (LIC) and, 218–219
  means for raising army and, 40–41
  mechanization of warfare and, 48–49
  military practitioner's perspective on, 189–190
  political science perspective on, 188
  sociology perspective on, 187–188
  strategy mismatch to, 213–214
  technology interrelatedness with, 20–21
  traditional strategic thought and, 37–49
  World War I and changes in approaches to, 47–49
Manpower issues
  active versus reserve forces and, 30–31
  combat versus support capabilities and, 30
  elite versus regular forces in, 31
  light versus heavy forces in, 29–30
  racial and ethnic composition and, 32–33
  unionization and, 33–34
  volunteerism versus conscription in, 31–32
  women in combat and, 32
Manpower policy, U.S.
  allied manpower and defense needs and, 60–61
  Army reserve shortages in, 56
  capital and labor in military applications and, 61
  Carter administration and, 54
  combat effectiveness and, 66–67
  comparative military power discussions in, 57
  conflict spectrum and, 74–76
  Congress and, 78–79, 83
  debates over military objectives and doctrine in, 69–70
  domestic reactions to, 79
  emphasis in, 54–55
  fighting ethic in soldiers and, 65–66, 73, 82–83
  future prospects in, 79–83, 141–182
  Joint Chiefs of Staff and, 189–190
  maneuver warfare and, 61–62
  manpower as output in, 55
  national objectives and, 70–71
  occupational orientation in volunteer system and, 66, 67
  options in, 89–139
  perceptions of constraints in, 62
  planning and, 71–73
  political system and, 77–79
  public support for, 188
  questions overlooked in, 81–82
  Reagan administration and, 54
  resource allocation process in, 57–58
  roundtable discussion of, 185–192
  socio-political dimensions of, 80–81
  strategic planning and, 53–62, 68–71, 72–73
  unconventional conflicts and, 69, 74, 76–77
  variables in planning in, 72
  Vietnam experience and, 78, 79, 84
  volunteer system and, 71
Manpower procurement, 91–136
  average annual requirements for male recruits in, 100

career/first-term ratio in, 112
career officers and, 127-131
civilian employees and, 111, 122
compensation and recruitment in, 108-110
computer use and, 101-102
conscription use in, 112-113, 134-135
current state of, 120-132
demographics and, 97-99
doctrine and impact on, 133-134
economic conditions and, 93, 96, 105-108, 121
junior officers and, 126-127
levels of, 120
limits on expansion of, 110-111
managerial elite and, 131-132
manpower management and, 110-112
national service concept and, 221-222
in next decade, 92, 97-108
noncommissioned officers (NCOs) in, 124-126
occupational mix and, 104-105
options in, 108-113
past debates over, 92-97
propensity of youth to enlist and, 96-97
proportion of males with technical skills in, 105, 106-107
proportion of qualified and available males required in, 97, 99
prospective military buildup in Reagan administration and, 99-100
quality of recruits and, 93, 94, 95, 96, 112, 120-121
readiness exercises and, 132
recruiting resource increases and, 109-110
recruitment goal changes and, 104-105
recruits in, 120-124
reform proposals for, 134-136
reserves and, 120, 132
retirement benefits and, 129-131, 135
survey of army officers and, 128-129
tank operation studies in, 103-104, 123-124
technology and skill mix and, 100, 101, 103, 104-105
two-career marriages and, 129
universal military obligation in, 187
women and, 110, 111, 122-123
see also Human resources
Manpower Requirements Report (Department of Defense), 55, 56, 57
Margiotta, Franklin, 55, 185, 186
Marine Corps
future conflict environments and, 154, 216
Lebanon incident with, 131, 144, 146, 150
technology modernization in, 103
Marriage patterns, and manpower procurement, 129
Martin, Michael L., 193
Mayaguez incident, 131
McLuhan, Marshall, 25, 26, 27
McPherson, Karen A., 8, 193-206
Mercenary army system, 41
Militia system, 41
Minority groups, *see* Racial composition of military
Morgenthau, Hans, 18
Moskos, Charles C., 66, 173
Myth, and military power, 27-29

Napoleon, 39
Napoleon III, 42
National Advisory Commission on Selective Service, 166
NATO
European attitudes toward, 28
future conflict environments and, 152
German membership in, 198
human resources planning and, 176, 178
manpower planning and, 73
racial composition of military and, 33
technology use and power in, 20
volunteer system and, 79
Navy, 147
civilian employees and, 111
proportion of males with technical skills in, 105, 106-107
technology modernization in, 102
Nazism, 201

New Deal, 164, 165, 166
New Manning System (Army), 159
Nicaragua, 148, 151
Noncommissioned officers (NCOs)
  manpower procurement and, 124–126
  quality of, 125–126
Normandy campaign, 65, 66
North Vietnam, 149
Nuclear weaponry
  deterrent power of, 145–146
  image in perception of, 24
  power and, 17
  Soviet balance and, 177
  strategy development and, 59

OCS, 126
Officer Personnel Management System (OPMS), 118
Operation Overlord, 65
Ostpolitik, 198

Palmer, General Bruce, 68
Panetta, Leon E., 113
Pay
  alternatives to manpower program and, 189
  human resources planning and, 172, 175–176
  manpower procurement and, 93–96, 108–110
  manpower strategy and, 58
  Reagan administration and, 109
Peace Corps, 222
Perceptions management
  active versus reserve forces and, 30–31
  combat versus support capabilities and, 30
  control move and, 23–24
  cross-national assessment of, 196–199
  elite versus regular forces in, 31
  European attitudes toward NATO and, 28
  Game of Machismo concept in, 26–27
  image importance in, 23–24
  light versus heavy forces in, 29–30
  manpower issues and, 29–34, 62
  policy imperative and, 25–26
  power assessment and, 24–25
  precepts in, 24
  racial and ethnic composition and, 32–33
  strategy and, 22–29
  symbolic interactionist component of, 24
  target group and image identification in, 26
  unionization and, 33–34
  volunteerism versus conscription in, 31–32
  women in combat and, 32
Perry, William J., 103
Persian Gulf, 149
Pirie, Robert B., Jr., 5, 53–62, 213, 220
Planning process, and manpower, 57
Political science perspective on manpower, 188
Political system
  cross-national assessment of, 199–202
  manpower planning and, 77–79, 80–81
Pol Pot, 148
Population changes
  cross-national assessment of, 202–203
  manpower procurement and, 97–99
  projections in, 97, 98
  proportion of qualified and available males required and, 97, 99
Power
  characteristics of, 16
  components of, 18–19
  conceptualizations of, 18
  definition of, 16
  force differentiated from, 18
  generalizability across range of threats of, 17–18
  importance of numbers in, 37
  interstate and interpersonal, 16–17
  myth and ritual and, 27–29
  nature of, 15–34
  perceptions and assessment of, 24–25
  strategy definition and, 15–16
  symbol related to human dimension of, 27
  technology and, 19–20
Private Sector Survey on Cost Control, *see* Grace Commission

Prussia, 38, 40
Pueblo incident, 144

Quality of recruits
  comparison of test scores of, 94, 95
  conscription and, 113
  in fiscal 1979, 120
  in fiscal 1984, 121
  future conflict environments and, 153-154
  manpower procurement debates and, 93, 94, 95, 96, 112
  military retirement system and, 131
  noncommissioned officers (NCOs) in, 125-126
  proportion of males with technical skills and, 105, 106-107
  readiness exercises and, 132
  skills qualification tests for, 120-121
  tank operation studies in, 103-104, 123-124

Racial composition of military
  human resources planning and, 175
  perceptions of, 32-33
Railway system, and troop transportation, 43
Reagan administration
  defense spending by, 19
  light versus heavy forces in, 29-30
  manpower procurement and, 110-111
  military pay and, 109
  prospective military buildup and, 99-100
  strategy formulation in, 54
  student aid programs and, 96
  technology modernization in, 100-101
Record, Jeffrey, 7, 143-160, 214, 217
Recruiters, increase in number of, 109-110
Recruitment, *see* Manpower procurement
Republican Party, 166, 174
Reserve forces
  perceptions and, 30-31
  procurement policy and, 120, 132
  strategy and shortages of, 56
Resources, and manpower, 15, 57

Retirement program
  cost of, 130
  manpower procurement and, 129-131
  reforms in, 135
Rifle volunteers, 46
Ritual, and military power, 27-29
Roberts, Lord, 46
Rogers, Bernard W., 56, 60
Rogers Plan, 157
Rose, Arnold, 24
ROTC, 126, 127, 136
Rudman, Warren, 111
Russia
  manpower sources for army of, 46
  see also Soviet Union

Sabrosky, Alan Ned, 9, 211-223
Salary, *see* Pay
Sandinistas, 148
Sarkesian, Sam C., 5, 65-85, 213, 214, 219, 223
Saudi Arabia, 149
Scandinavian countries, citizen forces in, 46
Schlieffen, 45
Scribner, Barry L., 104
Secrecy, and strategy development, 58-59
Segal, Mady W., 185, 187-188, 212
Sexual relationships among troops, 122-123
Sharpshooter movement, 46
Sinks, George W., 8, 185-192
Slave army, 41
Social change, and manpower policy, 66
Social dimension of manpower planning, 15, 80-81
Sociological perspective on manpower, 187-188
Soviet Union
  combat versus support capabilities and, 30
  conscription and, 114
  cross-national assessment with, 194, 196-197, 199-200, 202-203, 204, 213
  doctrine and military thought of, 21-22
  economic strength constraints in, 204

future conflict environments and, 149–150, 152
human resources planning and, 176–179
Iran and, 149–150
manpower emphasis of, 20
manpower policy overview in, 194
nuclear balance with, 147
political culture constraints in, 199–200
population constraints in, 202–203
threat perception in, 196–197
unionization of troops and, 33
volunteerism perceived by, 32
women in combat viewed, 32
Special operations units, 69, 82–83
Stockman, David, 130
Strategy
combat effectiveness and, 66–67
conflict spectrum and, 74–76
Congress and, 78–79, 83
debates over military objectives and doctrine in, 69–70
discipline and efficiency in, 38–39
discriminative principle in, 48
domestic reactions to, 79
fighting ethic in soldiers and, 65–66, 73, 82–83
future prospects in, 79–83
historic overview of approaches to, 37–49
importance of troop strength in, 37
manpower mismatch with, 213–214
manpower planning and, 37–49, 65–85
mechanization of warfare and, 48–49
national objectives and, 70–71
occupational orientation in volunteer system and, 66, 67
perceptions management and, 22–29
political system and, 77–79
power and definition of, 15–16
questions overlooked in, 81–82
socio-political dimensions of, 80–81
strategic planning and, 53–62, 68–71, 72–73
traditional approach to, 37–49
unconventional conflicts and, 69, 74, 76–77
U.S., *see* U.S. strategic planning
variables in planning in, 72

Vietnam experience and, 78, 79, 84
World War I and, 47–48
Student aid programs, 93, 96
Sun Tzu, 18, 22, 81
Support capabilities, perceptions of, 30
Symbol
meaning of, 24
military power related to, 27
myth and ritual and, 27–29
power assessment and, 24
Syria, 152

Tank operation studies, and manpower needs, 103–104, 123–124
Tax, Sol, 166–167
Taylor, William J., Jr., 3–9, 9, 211–223, 216
Technology
computer use and, 101–102
doctrine and importance of, 21–22
maintenance personnel and, 103–104
manpower analysis and, 13–14
manpower interrelatedness with, 20–21
manpower procurement issues and, 100–105, 219–220
occupational mix and, 104
personnel for operation of, 21
power and, 19–20
proportion of males with technical skills in, 105, 106–107
Reagan administration modernization of, 100–101
military retirement system and, 131
skill mix needs and, 100, 101, 103
Soviet Union and, 177–178
strategy-manpower interface and, 211–212
tanks and, 103–104
Tennessee Valley Authority, 165
Tennyson, Alfred, 46
Territorial Force (Great Britain), 46
Terrorism, 148, 151
Third World
cross-national assessment with, 195, 198–199, 201–202, 203, 205, 213
economic strength constraints in, 205
future conflict environments and, 143, 147–148

manpower policy overview in, 195
political culture constraints in, 201–202
population constraints in, 203
threat assessment and, 190, 198–199
Torricelli, Robert G., 113, 174
Transportation of troops, 43
Turkey, 149

Unconventional conflicts, 144
future conflict environments and, 151
manpower planning and, 69, 74, 76–77
Unemployment
human resources planning and, 166
manpower procurement and, 93, 121, 124
Uniforms, and ritual and ceremony, 27–29
Unionization, perceptions of, 33–34
U.S. Central Command (CENTCOM), 149
U.S. strategic planning, 53–62
allied manpower and defense and, 60–61
ambiguities and uncertainties in, 59
army-people relationship and, 78
assigning priorities to probable missions and, 215–216
capital and labor in military applications and, 61
Carter administration and, 54
combat effectiveness and, 66–67
comparative military power and structure analysis in, 57
conflict spectrum and, 74–76
Congress and, 78–79, 83
debates over military objectives and doctrine in, 69–70
domestic reactions to, 79
emphasis on manpower in, 54–55
fighting ethic in soldiers and, 65–66, 73, 82–83
future prospects in, 79–83
individual ready reserve (IRR) in, 62
maintaining public support for, 188
maneuver warfare emphasis in, 61–62
manpower as output in, 55
manpower planning and, 72–73
national objectives and, 70–71
nature of development of, 58–59
occupational orientation in volunteer system and, 66, 67
perceptions of constraints in, 62
political system and, 77–79
questions overlooked in, 81–82
Reagan administration and, 54
reserve forces shortages and, 56
resource allocation process in, 57–58
role of manpower considerations in, 57–60
roundtable discussion of, 185-192
secrecy in, 58–59
socio-political dimensions of, 80–81
terminology used in, 59–60, 68
unconventional conflicts and, 69, 74, 76–77
universal military obligation in, 187
variables in planning in, 72
Vietnam experience and, 78, 79, 84

van Creveld, Martin, 170
Vessey, General Jon W., Jr., 174
Vietnam War, 150, 155, 166, 218
army-people relationship and, 78
career officers and, 127–128
command authority in, 157
fighting instincts in soldiers and, 66
human resources planning and, 167–170
manpower planning and experience in, 76–77, 79, 81, 84
racial composition of troops in, 175
volunteer force and, 219
Voltaire, 37, 211
Volunteer army
alternatives to, 71
comparison of recruit test scores in, 93, 94, 95
demographics and, 97–99, 219
early European use of, 46–47
economic conditions and, 105–108, 191
enlistment terms in, 156
future conflict environments and, 153–154
future threats to, 92
human resources planning and, 166–167, 173–174
low-intensity conflict (LIC) and, 218

manpower availability and, 20
manpower debate in, 71
manpower management and, 110–112
national service concept and, 221–222
occupational orientation of, 66, 67
perceptions of, 31–32
propensity of youth to enlist and, 96–97
racial and ethnic composition of, 33
Soviet perception of, 32
Vietnam syndrome and, 219
women and, 110

Warrior principle, 41
Wars of national liberation, 144
Wartime Manpower Planning System (WARMAPS), 213
Watergate affair, 79
Weigley, Russell, 69
Weimar Republic, 201
Weinberger, Caspar W., 54, 174
Wells, Samuel F., 193
Wesbrook, Stephen D., 73
Western European Union (WEU) treaty, 195
West Germany
  cross-national assessment with, 194–195, 198, 200–201, 203, 204, 213
  economic strength constraints in, 204
  manpower policy overview in, 194–195, 205
  political culture constraints in, 200–201
  population constraints in, 203
  racial composition of military and, 33
  threat perception in, 198
Weyand, Genral Fred C., 66, 78, 79
William I of Prussia, 41
Women troops
  combat and, 123
  human resources planning and, 175
  manpower procurement issues and, 110, 111, 122–123
  perceptions of, 32
  sexual relationships and, 122–123
World War I, 47–49, 145, 186, 212
World War II, 145, 156, 166
  manpower strategy in, 48, 186
  Normandy campaign in, 65, 66

Youth Attitude Tracking Study, 96–97

Zoll, Ralf, 193

ABOUT THE EDITORS

Gregory D. Foster is a Vienna, Virginia-based consultant on international security affairs and an adjunct faculty member at Johns Hopkins University and the University of Southern California.

Alan Ned Sabrosky is Director of Studies, Strategic Studies Institute, and holder of the General of the Army Douglas MacArthur Chair of Research at the U.S. Army War College.

William J. Taylor, Jr., is Executive Vice President and Chief Operating Officer of the Georgetown University Center for Strategic and International Studies.

ABOUT THE CONTRIBUTORS

Martin Binkin is a Senior Fellow in the Foreign Policy Studies Program at the Brookings Institution.

William L. Hauser is Director of Career Development for Pfizer Inc.

Irving Louis Horowitz is Hannah Arendt Professor of Sociology and Political Science at Rutgers University and Editor in Chief of *Transaction/SOCIETY*.

John Keegan is Military Correspondent for the London *Daily Telegraph*.

Karen A. McPherson is Director of Information Services at the BDM Corporation, McLean, Virginia.

Robert B. Pirie, Jr., is Assistant Vice President of Program Development and Review at the Institute for Defense Analyses and a former Assistant Secretary of Defense (Manpower, Reserve Affairs and Logistics).

Jeffrey Record is Adjunct Professor of Military History at Georgetown University and Military Commentator for *The Baltimore Sun*.

Sam C. Sarkesian is Professor of Political Science at Loyola University of Chicago and Chairman of the Inter-University Seminar on Armed Forces and Society.

George W. Sinks is a Fellow at the Georgetown University Center for Strategic and International Studies.

RAYMOND H. FOGLER LIBRARY
DATE DUE

BOOKS ARE SUBJECT TO
CALL AFTER TWO WEEKS

3 0 1987